Virtual Reality and Animation for MATLAB® and Simulink® Users

Nassim Khaled

Virtual Reality and Animation for MATLAB® and Simulink® Users

Visualization of Dynamic Models and Control Simulations

 Springer

Nassim Khaled
Cummins Technical Center
1900 Mckinley Avenue
Columbus, IN 47201
USA
nassim.khaled@gmail.com

ISBN 978-1-4471-2329-3   e-ISBN 978-1-4471-2330-9
DOI 10.1007/978-1-4471-2330-9
Springer London Dordrecht Heidelberg New York

British Library Cataloguing in Publication Data
A catalogue record for this book is available from the British Library

Library of Congress Control Number: 2011943558

Printed on acid-free paper

Springer is part of Springer Science+Business Media (www.springer.com)

Contents

Chapter 1
Introduction to Virtual Reality in MATLAB® and Simulink® Environment

1.1 Overview of the Book

Whether the reader wants to model the dynamics of a simple pendulum or design a controller for a robot arm, it is always useful to be able to animate the dynamic model. Animation of a dynamic model enables one to observe all the degrees of freedom of a system at once, rather than looking at 2D and 3D plots of simulated data. Such animations serve as a debugging tool for developing the dynamic model in hand. It can also help in preliminary assessment of the performance of the controller. In addition, animations can help students and less-technical audience to understand the end results of a dynamic model without going into lengthy mathematical derivations.

Two of the top leading tools in dynamic modeling and control development are MATLAB® and Simulink®. Researchers and the industry have been increasingly using MATLAB® and Simulink® in the past two decades. One would be surprised to know that many of M-script file users are not comfortable using Simulink® and vice versa. Thereby the author has decided in this book to address both users separately. For both sets of users, the first step to create a virtual reality animation is to draw the object(s) (along with the virtual scene) in V-Realm Builder, and the second is to use MATLAB® M-script or Simulink® model to vary the states or properties of the object(s) and/or the scene based on a set of differential equations, prescribed trajectories, or empirical formulas.

The main theme of this book along with the supplementary electronic material is to "learn through application." Step-by-step examples in each chapter with illustrative screen shots will guide the reader through this learning experience. Furthermore, users can download the virtual scenes, the movie recordings for the animations, and the M-script files (or the Simulink® models) for the examples included in each chapter from Springer's web site by following the link http://extras.springer.com/

Matlab® and Simulink® are registered trademark of The Mathworks, Inc.

N. Khaled, *Virtual Reality and Animation for MATLAB® and Simulink® Users:*
Visualization of Dynamic Models and Control Simulations,
DOI 10.1007/978-1-4471-2330-9_1, © Springer-Verlag London Limited 2012

and searching for the ISBN of the book. These downloadable materials should help users understand the end result of each example, explore the models, and debug their own models.

The author tried to keep the chapters as independent as possible. If the reader is not interested in becoming an expert in virtual reality, this introduction will help him/her skip to his/her desired application. It is still strongly advised to go through all the chapters in order to gain a good level of experience in virtual reality.

Chapter 1 of this book is a short introduction to the book and its contents. Chapter 2 describes the main tool, V-Realm Builder, which is used to set up and draw the virtual reality objects and the virtual environment. Chapters 3, 4, 5, 6, and 7 are intended for M-script file users. They contain step-by-step examples with detailed screen shots on how to create an object in V-Realm Builder and how to write an M-script file to link the dynamic model (that is governed by equation(s) of motion, empirical formula, or prescribed trajectory) to the virtual scene in order to animate the object(s) and/or the surrounding scene. Chapters 3, 4, 5, 6, and 7 are sorted in ascending level of difficulty. Each chapter includes the utilization of new properties and features of V-Realm Builder. Chapters 3, 4, 5, 6, and 7 contain the following examples, respectively:

- The translating cube in addition to an application problem of the collision of two objects
- The mass-spring-damper oscillations in addition to an application problem of two masses connected by a spring
- The crank-slider mechanism of a piston in addition to an application problem of PID control of a ball on a plate
- Car animation with joystick control in addition to an application problem of fuzzy logic control of the speed of the car
- Animation of a ship moving across waves in addition to an application problem of fuzzy logic control of the heading of the ship

Chapters 8, 9, 10, 11, and 12 contain a step-by-step guide with detailed screen shots to animate the virtual worlds constructed in Chaps. 3, 4, 5, 6, and 7 using Simulink® in addition to the application problems presented. Finally, a short index to the supplementary material of the chapters is covered at the end of the book. The supplementary material includes movie recordings for all the problems covered in the book and M-script files and Simulink® models that were used to animate the virtual systems in addition to all the virtual scenes that were constructed in Chaps. 3, 4, 5, 6, and 7. The supplementary material can be downloaded from Springer's web site, http://extras.springer.com/, by searching for the ISBN of the book.

Few comments regarding Matlab® version and the operating system:

- the codes of this book have been developed using Matlab® 2008a and windows 7 operating system and checked for compatibility using Matlab® 2011b.
- as of Matlab® 2009a, **Virtual Reality Toolbox** has been renamed **Simulink 3D Animation** thus users who will be using Matlab® 2009a or later should know that the only difference, in the context of this book, is the name of the toolbox.

Chapter 2
V-Realm Builder©

2.1 Introduction

Section 2.2 of this chapter gives the user a brief outline of V-Realm Builder Software© which will be used in Chaps. 3, 4, 5, 6, and 7 to construct the virtual objects and scenes.

Section 2.3 guides the user through the installation process of the V-Realm Builder Software©.

Section 2.4 helps the user start V-Realm Builder Software©.

2.2 What Is V-Realm Builder?

V-Realm Builder (which in the context of this book will sometimes be referred to as VRML) is a software that is used to design virtual worlds and draw/import 3D virtual objects. After the user builds his/her virtual world, he/she can manipulate the objects and the virtual scene through MATLAB® commands and Simulink® models to animate the scene. Virtual 3D objects' properties such as translation, rotation, scale, and color can be changed through MATLAB® commands and Simulink® models.

2.3 Installing V-Realm Builder from MATLAB®

The user needs to purchase the MATLAB®'s Virtual Reality Toolbox® to have access to V-Realm Builder. To check if Virtual Reality Toolbox® is installed, the following command should be typed in the MATLAB® command window:

vrinstall -check

V-Realm Builder 2.0 is copyrighted, 1996-1997, by Integrated Data Systems, Inc. Matlab® is a registered trademark of The Mathworks, Inc.

N. Khaled, *Virtual Reality and Animation for MATLAB® and Simulink® Users:*
Visualization of Dynamic Models and Control Simulations,
DOI 10.1007/978-1-4471-2330-9_2, © Springer-Verlag London Limited 2012

If the V-Realm Builder is installed correctly, the following statement will be returned in the MATLAB® command window:

VRML viewer: installed
VRML editor: installed

If VRML viewer or editor is not installed, you can reinstall them both by typing the following in the MATLAB® command window:

vrinstall -install

MATLAB® will display the following message in the command window:

Installing blaxxun Contact viewer…
Do you want to use OpenGL or Direct3D acceleration? (o/d)

The user should type **o** (or **d**) and hit **Enter** or **Return** on the keyboard and follow the screen instructions to finish the installation.
If MATLAB® command window returns the following:

??? Undefined command/function 'vrinstall.',

Then the user does not have the Virtual Reality Toolbox®. It can be purchased by contacting a sales representative for MathWorks by the following link: http://mathworks.com/.

2.4   Starting V-Realm Builder

In order to start V-Realm Builder, the user has to go to the location where MATLAB® was installed and go to the following subdirectory location:

{Matlab installation folder}\toolbox\vr\vrealm\program

In other words, if MATLAB® was installed in *C:\Matlab*, then the user can access V-Realm Builder by going to the following directory (inside the rectangle in Fig. 2.1):

C:\Matlab\toolbox\vr\vrealm\program

When the user goes to the right subdirectory, he/she will see V-Realm Builder executable, which is named *vrbuild2* (inside the ellipse in Fig. 2.1).
If the user is going to use V-Realm Builder frequently, it is easier to create a shortcut on the desktop or any other convenient location for *vrbuild2* rather than accessing the directory every time. To launch V-Realm Builder, the user should double-click *vrbuild2*, and the main window of V-Realm Builder will open (Fig. 2.2).

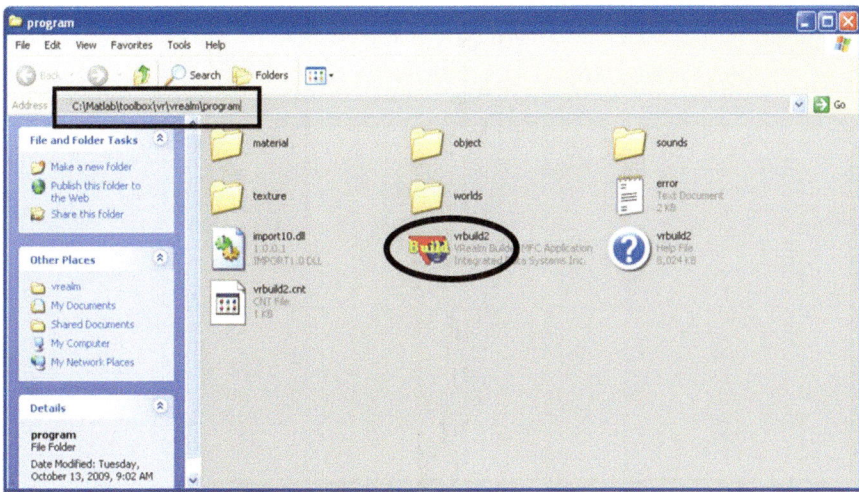

Fig. 2.1 Directory location of V-Realm Builder

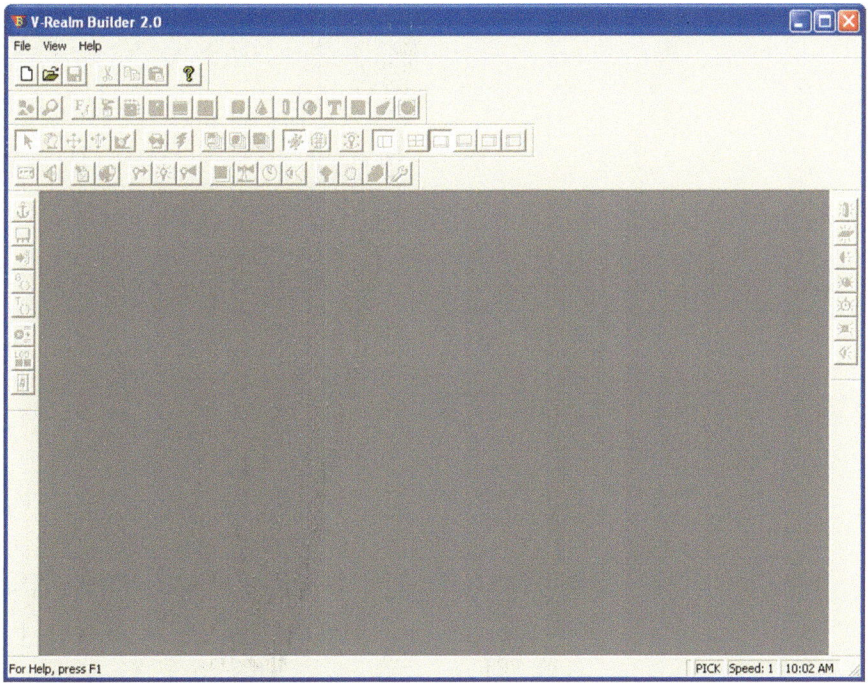

Fig. 2.2 Main window of V-Realm Builder

Chapter 3
The Translating Cube

3.1 Introduction

The purpose of this chapter is to introduce MATLAB® users that are interested in animating their physical system in a virtual reality environment to the virtual reality tool, V-Realm Builder (VRML). This chapter will walk the new user through a complete exercise on how to animate a physical problem governed by an equation(s) of motion in VRML environment using M-script commands.

In Sect. 3.2, a simple dynamics problem of a cube that is subjected to an external force will be presented. Newton's second law will be used to derive the equation of motion.

In Sect. 3.3, step-by-step instructions will guide the reader through drawing the cube in VRML and changing its name, color, and size. A simple background will also be added to the scene.

The x-, y-, and z-axes convention of VRML will be briefly explained in Sect. 3.4. In addition, a simple familiarization experiment will be described to enable the reader to better understand the coordinate system of VRML.

In Sect. 3.5, an M-script file that numerically computes the solution of the differential equation (derived from Newton's second law) will be developed.

To animate the virtual scene, an M-script file that uses the numerical solution from Sect. 3.5 will be described in Sect. 3.6. Furthermore, a procedure to save the animation as a movie file will be included.

The chapter concludes by an application problem of two colliding masses in Sect. 3.7.

The electronic version of all the M-script files and VRML models in addition to the recorded movies for the cube and the collision problem can be downloaded from Springer's web site http://extras.springer.com/.

Matlab® is a registered trademark of The Mathworks, Inc.

N. Khaled, *Virtual Reality and Animation for MATLAB® and Simulink® Users: Visualization of Dynamic Models and Control Simulations*, DOI 10.1007/978-1-4471-2330-9_3, © Springer-Verlag London Limited 2012

3.2   Cube Problem

Given a cube of mass $m = 1(\text{kg})$ and of dimensions $0.5 \times 0.5 \times 0.5 (\text{m}^3)$. The cube is subjected to the following external force: $F_z = -1\,(\text{N})$.

Using Newton's second law, the following second-order differential equation for the z-displacement (check Sect. 3.4 for coordinate's convention) of the cube can be developed:

$$m\ddot{z}(t) = F_z \Rightarrow \ddot{z}(t) = \frac{F_z}{m} \tag{3.1}$$

The purpose of this exercise is to animate the solution of this equation for zero initial conditions.

3.3   Creating the Virtual Scene for the Cube

To create the virtual scene for the cube, follow the following steps:

1. Open the V-Realm Builder and click on **File > New** to create a new virtual world.
2. Click on the **Insert Background** button (Fig. 3.1) to create a background for the scene.
3. Click on the **Insert Box** (Fig. 3.2) to create the cube. In the main view window, you should see a cube with a green and blue background behind it. Notice that on the left side of the main view window, some text and yellow icons have appeared, all under the title **Transform** (Fig. 3.3).
4. Click once on the title **Transform** to select it. Click another time on **Transform** and change the text to a meaningful name such as **Cube** (Fig. 3.4). Do not double-click on **Transform**; you should wait at least 1 s after clicking the first time. The purpose behind changing the name from **Transform** to **Cube** is to allow MATLAB® to uniquely identify the object **Cube**. In other words, suppose you wanted to have two cubes instead of one, then both of the cubes will have the name **Transform**, and MATLAB® will not be able to differentiate between both objects. In this case, you better name one cube as **Cube1** and the other as **Cube2**.
5. Click on the **Color Mode** button (Fig. 3.5).
6. Double-click on the **Diffusive Color** gray area (Fig. 3.6).
7. Click on the blue color or any other color you prefer (Fig. 3.7). Press OK when done.
8. Click on any face of the cube, and the whole object will turn blue (Fig. 3.8). Press Close when done.
9. By default, the size of the cube is $2 \times 2 \times 2$ and its center is positioned at the origin $(0,0,0)$. To change its size, click the $+$ sign to the left of **Box** and then double-click on **size** (Fig. 3.9). Tick the x-axis box (Fig. 3.10) and change

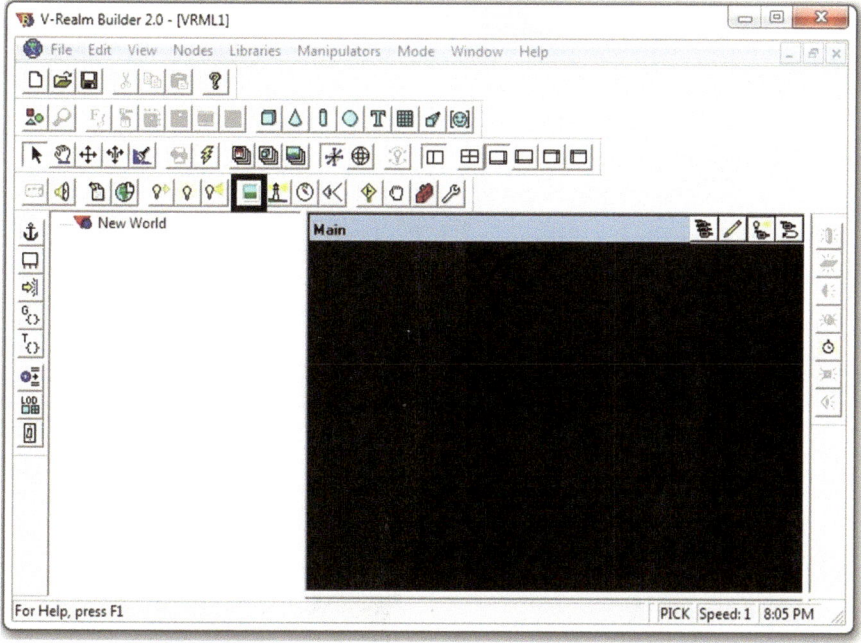

Fig. 3.1   Insert Background button is shown inside the rectangle

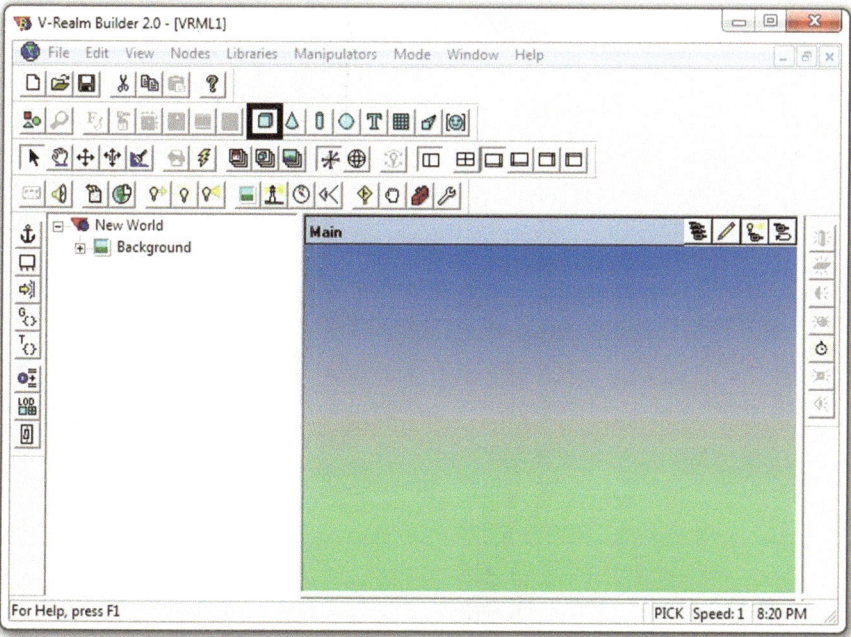

Fig. 3.2   Insert Box button is shown inside the inside the rectangle

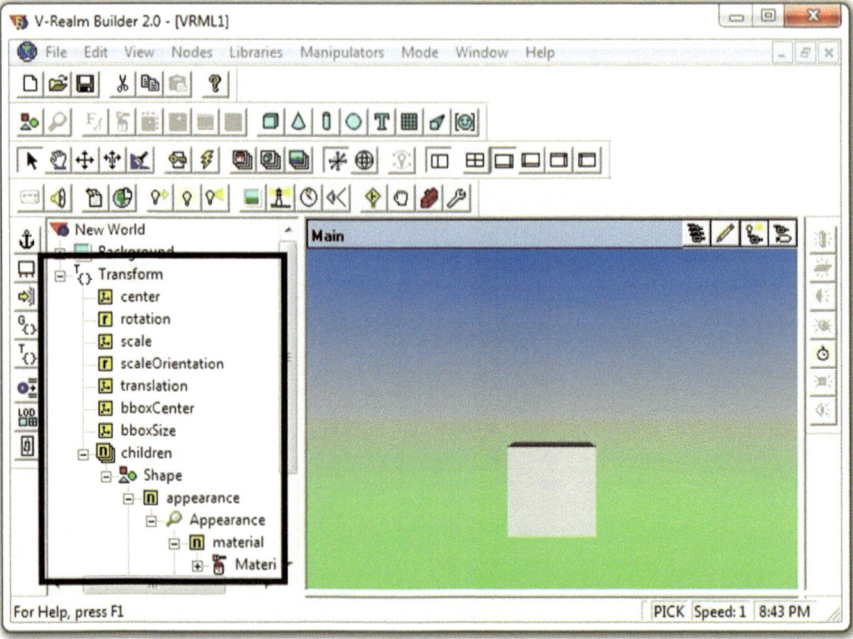

Fig. 3.3   The text and the yellow icons appear under the title Transform

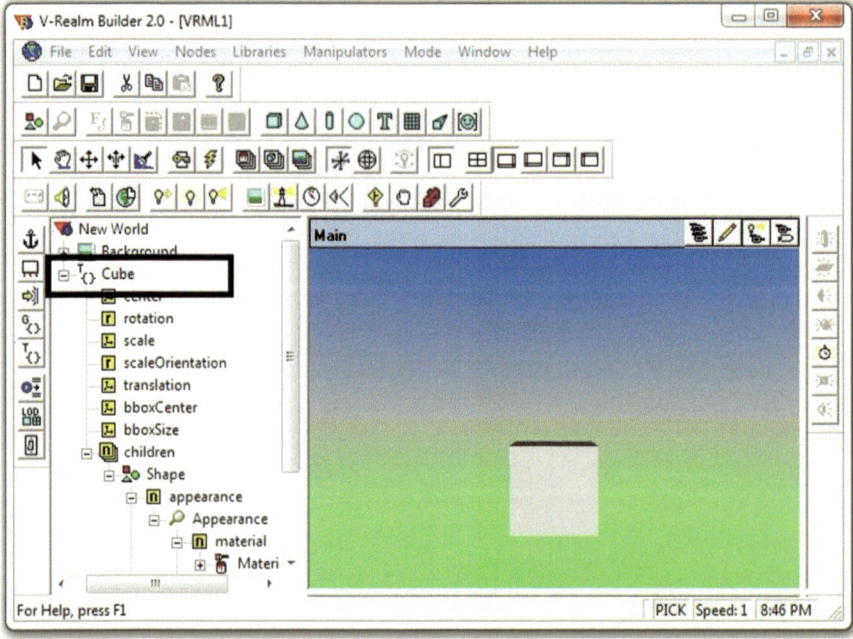

Fig. 3.4   Change the name of the cube from Transform to Cube

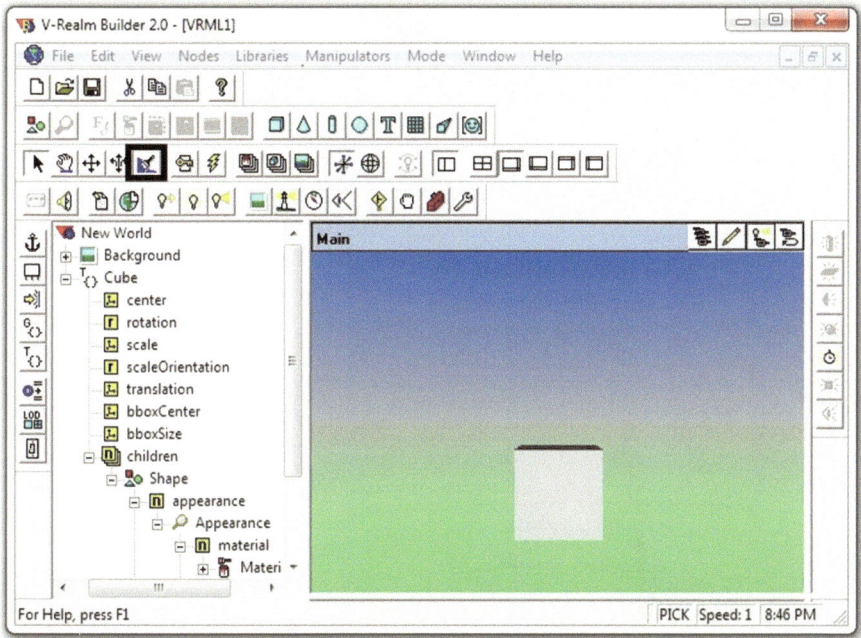

Fig. 3.5 Color Mode button is shown inside the rectangle

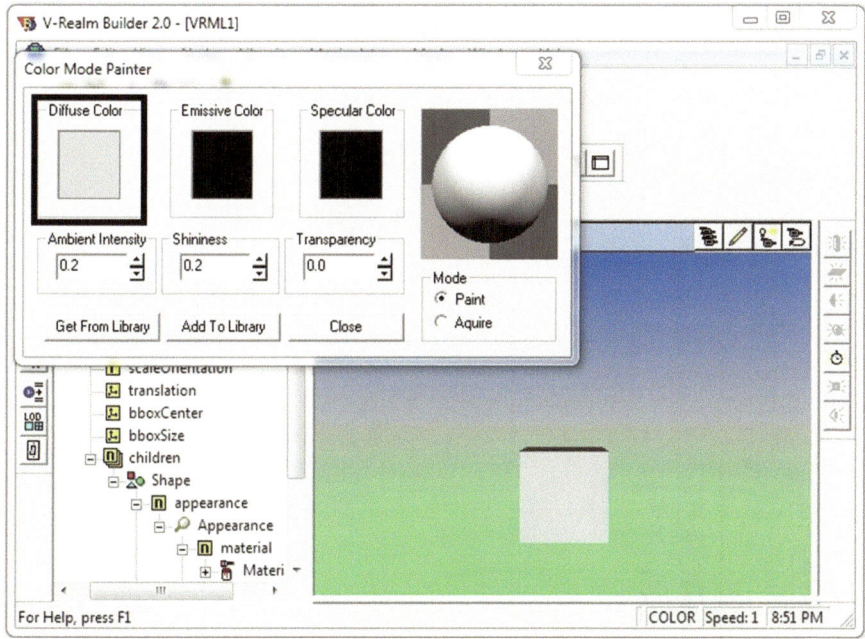

Fig. 3.6 Diffusive Color gray area inside the rectangle

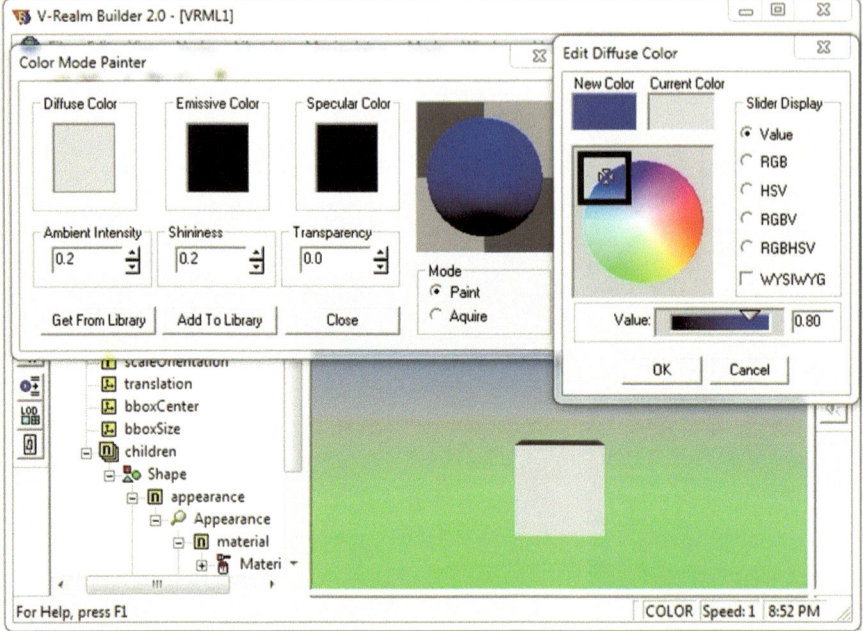

Fig. 3.7 Click on the blue area to choose the blue color

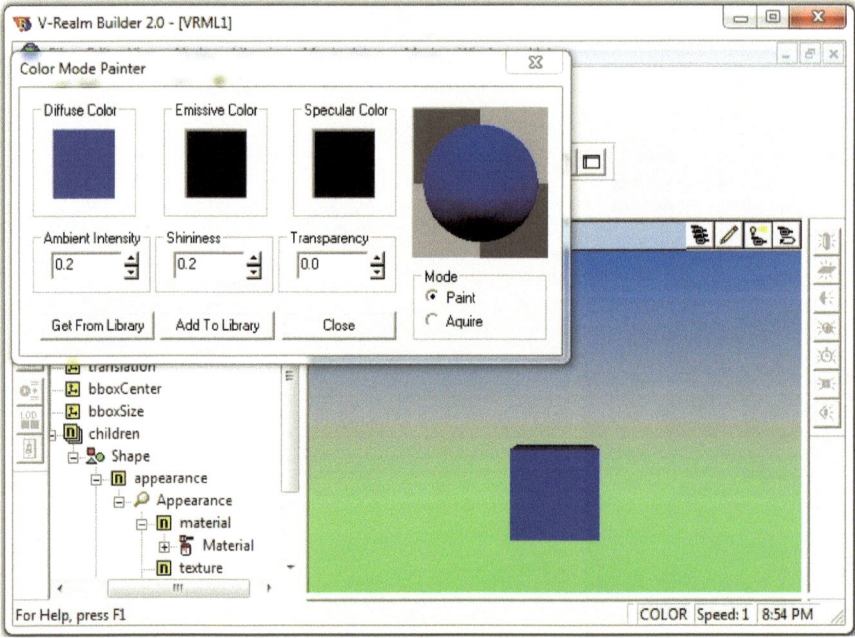

Fig. 3.8 Click on any face of the cube

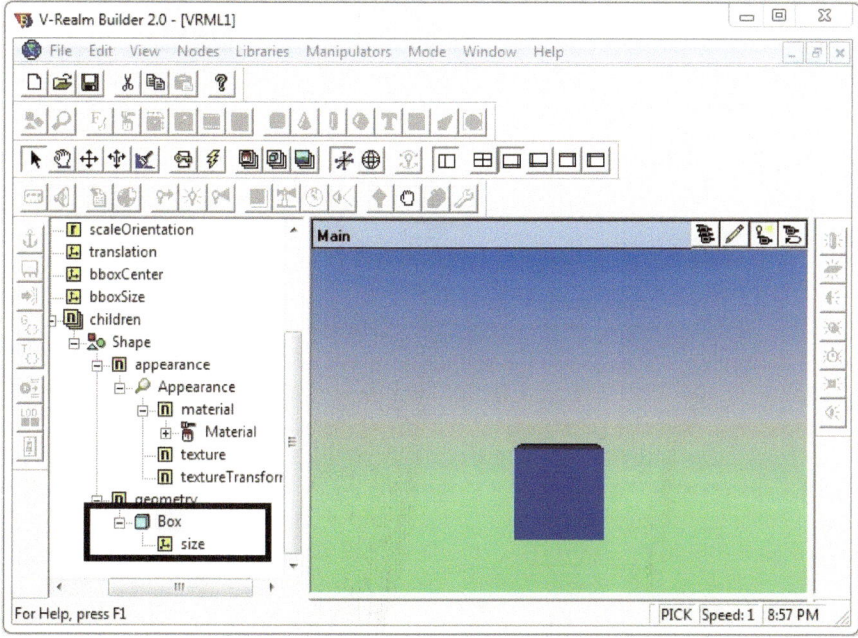

Fig. 3.9 The size property is shown inside the rectangle (under Box)

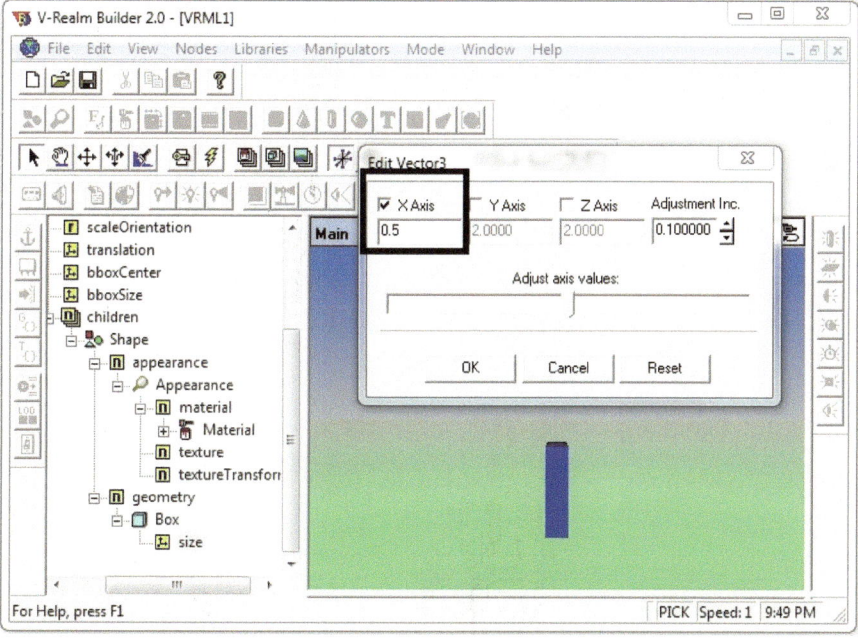

Fig. 3.10 Tick the *x*-axis box by clicking on the box in the rectangle

the value to 0.5. Similarly, tick the y- and z-axes boxes, and change their respective values to 0.5. Press OK when done. Notice the change in the size of the cube.

10. Click on **File > Save as**, and save it in MATLAB® working directory. Name the file ***Cube_Virtual***. The extension of your virtual reality file will be ***.wrl***.

3.4   Cartesian Coordinates for the VRML

The plane of the screen is of two dimensions. In VRML, the x-axis is the horizontal line of the screen and points to the right. The y-axis is vertical and points upward. As for the z-axis, it is perpendicular to the plane of the screen, and it is pointing toward the user (Fig. 3.11).

To experiment with the convention of the Cartesian coordinates of VRML, double-click the **translation** property of **Cube** (Fig. 3.12). Tick the x-axis box and then use the slider bar to vary the x-value. Notice that the positive values of x are moving the box in the right direction. Tick the x-axis box to unselect it and then tick the

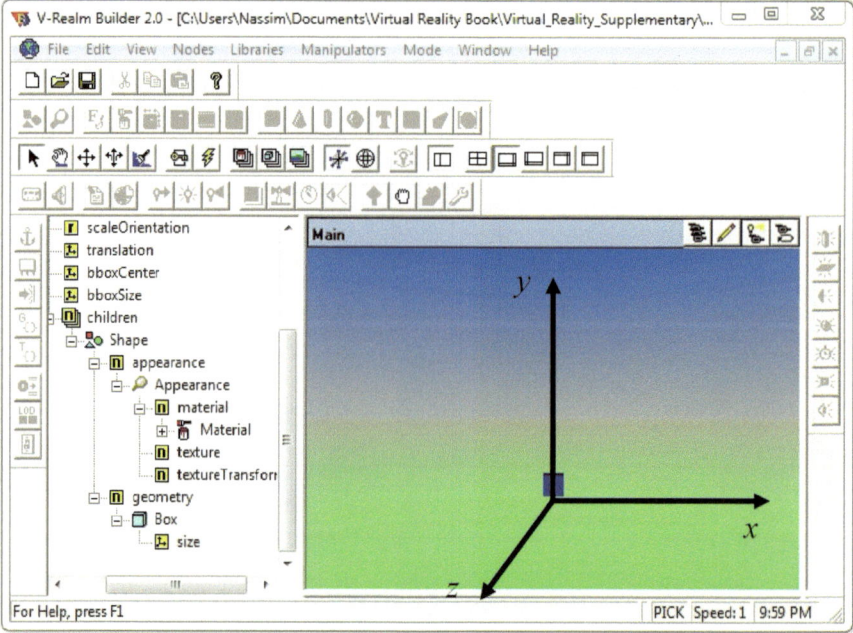

Fig. 3.11 Cartesian coordinates for VRML

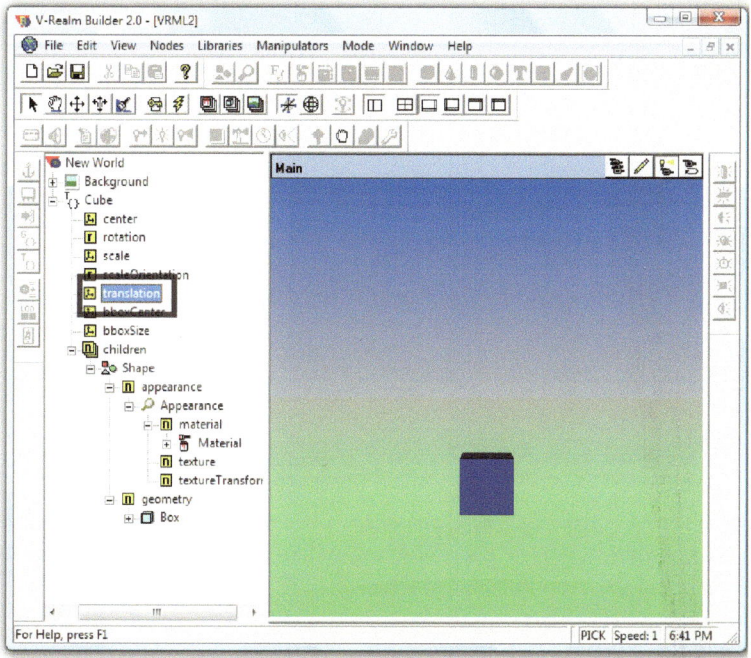

Fig. 3.12 The translation property is shown inside the rectangle

y-axis box. Use the slider bar to vary the *y*-value. Notice that the positive values of *y* are moving the box upward. Tick the *y*-axis box to unselect it, and then tick the *z*-axis box. Use the slider bar to vary the *z*-value. Notice that that box is moving closer for positive values of *z*.

3.5 M-script File for Solving the Differential Equation

To numerically integrate for the solution of the differential equation given in Eq. 3.1, use the **ode45** function in MATLAB®. For more information about this function, type in the MATLAB® command window: **help ode45**.

The following M-script file, ***cube_solver.m***, will numerically integrate the differential equation given in Eq. 3.1. Some comments have been added to briefly explain the M-script file. The M-script can be found in Chap. 3 folder of the extra material of the book. The folder can be downloaded by following http://extras.springer.com/ and then searching for the book by entering the ISBN or the title.

cube_solver.m

```
%This is the m-script file for the example Translating Cube:
% Initial time:
t0=0;
% Final time:
tf=10;
% Number of samples (for numerical integration of the solution)
N=50;
% Initial z position:
z_0=0;
% Initial z velocity:
z_dot_0=0;
% Compute the solution of the differential equation using ode45
[t,z]=ode45(@myfunc,[t0:(tf-t0)/N:tf],[z_0 z_dot_0]);
%Plot the displacement
plot(t,z(:,1));
```

The above M-script file, *cube_solver.m*, calls *myfunc.m* which computes the derivative vector $[\dot{z}(t)\,\ddot{z}(t)]$. The following is the script for *myfunc.m* which can also be found in the extra material of Chap. 3:

myfunc.m

```
function dzdt=myfunc(t,z)
%Mass
m=1;
%Force
F=-1;
%z_dot
dzdt_1=z(2);
%z_dot_dot
dzdt_2=F/m;
%Derivative vector
dzdt=[dzdt_1;dzdt_2];
```

3.6 M-script File for Animating the Virtual Scene

In what follows, the virtual scene, *Cube_Virtual.wrl*, will be animated based on the solution given by *cube_solver.m* (the reader should run *cube_solver.m* before running *cube_animate.m*), and the animation will be recorded into a movie file, *Cube_Movie.avi*. The only parameter that will be changed in the virtual world is the **translation** property of the cube.

A first look at *cube_animate.m* script might seem overwhelming and full of commands; thereby the reader is advised to start writing his/her own code by modifying an existing one instead of starting from scratch every time.

The script that comes before the for-loop consists of two parts. The first one (refer to Part 1 of *cube_animate.m*) deals with opening and viewing the virtual world. The second (refer to Part 2 of *cube_animate.m*) deals with setting the recording of the animation to a ***.avi** file. In this example, the movie will be saved as ***Cube_Movie.avi***.

As was mentioned before in this section, the only parameter that is being varied in this virtual world is the **translation** property of the cube. Going back to Fig. 3.12, the object that was named Cube has a **translation** property. The created virtual world is named "**world**" in Part 1 of the code:

world = vrworld('Cube_Virtual');

To change the **translation** property, assign a new value for the **translation** by writing world.Cube.translation = [0 0 z(Index,1)];

After you run ***cube_animate.m***, a video file ***Cube_Movie.avi*** should be created in the same directory as the MATLAB® file. Try playing the movie to see a playback of the animation. To increase or decrease the duration of the movie, change the number of samples used for the numerical integration of the differential Eq. 3.1.

cube_animate.m

```
%%%% Part 1
% Create a virtual world associated with Cube_Virtual.wrl file
world = vrworld('Cube_Virtual');
%Open the virtual world
open(world);
%Draw the virtual world
fig = view(world, '-internal');
%
%
%
%%%% Part 2
%Define the interval time to record the animation
set(world, 'RecordInterval', [0 tf]);
%Define the name of the movie file to be Cube_Movie.avi
set(fig, 'Record2DFileName','Cube_Movie.avi');
% Enable 2D recording
set(fig, 'Record2D', 'on');
%Compression quality of the movie
set(fig, 'Record2DCompressQuality', 100);
%Define recording mode to be scheduled
set(world, 'RecordMode', 'scheduled');
%Initiate the counter
```

```
Index=0;
%Sweep the time interval
for t=t0:(tf-t0)/N:tf
    %Increment the counter
    Index=Index+1;
    %Change the translation property of the object Cube that was defined
    %in Cube_Virtual.wrl to the value computed by ode45
    world.Cube.translation=[0 0 z(Index,1)];
    %Change the time of the virtual world to match the time that was
    %given by ode45
    set(world,'Time', t);
    %Update the virtual reality scene
    vrdrawnow;
end
%Close the virtual world
close(world);

delete(world);
```

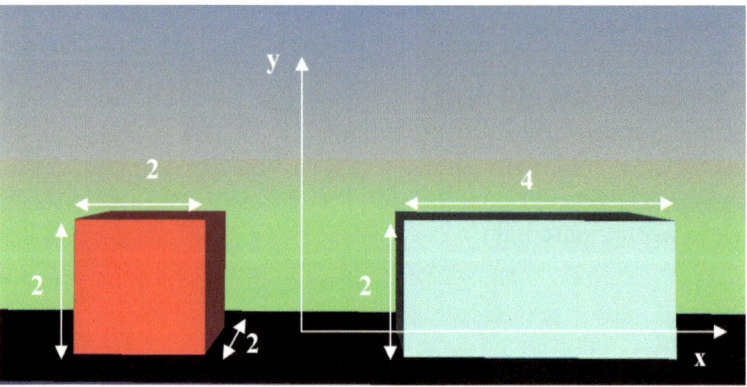

Fig. 3.13 Dimensions for the boxes

3.7 Application Problem

Given two boxes of mass $m_{B1} = 1\,(\text{kg})$ and $m_{B2} = 2\,(\text{kg})$ and their dimensions are $2 \times 2 \times 2\,(\text{m}^3)$ and $4 \times 2 \times 2\,(\text{m}^3)$, respectively (Fig. 3.13). The initial velocity of box 1 (the smaller box) in the x-direction is $1\,(\text{m}/\text{s})$, while box 2 is initially at rest. The initial position of box 1 is at $(-3,1,0)$, while box 2 is initially at $(3,1,0)$. Solve the following two parts (the solution is included in the electronic material for Chap. 3):

Part 1

Assuming no friction with the ground, and using a coefficient of restitution of $e = 0.9$ (Hibbler 1986):

1. Reconstruct Fig. 3.13 in VRML assuming that the bottom faces of the boxes are located at $y = 0$ and that the two boxes are aligned horizontally. Add an additional box component to represent the ground where the top face is at $y = 0$.
2. Develop the equations of motion of boxes 1 and 2 before and after the collision (Hibbler 1986).
3. Develop the M-script file to compute the numerical solution of the equations of motion and animate the VRML model. Assume that the total time of simulation is 10 s and that the step size is 0.01 s.
4. Plot the x-positions and x-velocities of boxes 1 and 2 as function of time. Plot the combined energy (potential and kinetic) of the system as function of time.

Part 2

Assume that the friction with the ground is opposing the motion and takes the form $-\mu \times (normal\ force)$. μ is the friction coefficient (Hibbler 1986) and is equal to 0.005 in this part, and *normal force* is equal to $m \times g$, where g is the gravitational acceleration and is equal to 9.81 m/s^2.

1. Develop the equations of motion of boxes 1 and 2 before and after the collision.
2. Develop the M-script file to compute the numerical solution of the equations of motion and animate the VRML model. Assume that the total time of simulation is 10 s and that the step size is 0.01 s.
3. Plot the x-positions and x-velocities of boxes 1 and 2 as function of time. Plot the combined energy (potential and kinetic) of the system as function of time.
4. Compare the energy from Parts 1 and 2. What do you notice?

The solution of the problem can be downloaded from Springer's web site http://extras.springer.com/. The files can be found in folder /Chapter 3/Application Problem.

Reference

Book

Hibbler RC (1986) Engineering mechanics: dynamics, 4th edn. McMillan, New York

Chapter 4
Mass-Spring-Damper Oscillations

4.1 Introduction

The purpose of this chapter is to familiarize MATLAB® users to the process of animating a physical system that has more than one component in a virtual reality environment. This chapter will walk the reader through a complete exercise on how to animate a physical problem that consists of two or more objects governed by an equation(s) of motion in VRML environment using M-script commands.

In Sect. 4.2, a mass-spring-damper problem will be presented where the mass is attached to a spring and subjected to an initial displacement. Newton's second law will be used to derive the equation of motion.

In Sect. 4.3, step-by-step instructions will guide the reader through drawing the virtual scene in VRML. The components of the system that will be drawn are the spring, the mass, the floor, and the wall.

In Sect. 4.4, the mass will be colored, and a new texture will be added to the floor and wall.

Section 4.5 will include a description for the M-script files that will numerically integrate for the equation of motion.

Section 4.6 will include a description of the M-script file that will animate the virtual scene based on the numerical solution from the previous section. In addition, the script will record the animation into a video file.

The chapter concludes by an application problem of two masses connected by a spring. The problem will include constructing the virtual scene in addition to animating the physical system based on a set of governing differential equations.

The electronic version of all the M-script files and VRML models in addition to the recorded movie for the mass-spring-damper and the oscillations of the two masses can be downloaded from Springer's web site http://extras.springer.com/.

Matlab® is a registered trademark of The Mathworks, Inc.

N. Khaled, *Virtual Reality and Animation for MATLAB® and Simulink® Users: Visualization of Dynamic Models and Control Simulations*,
DOI 10.1007/978-1-4471-2330-9_4, © Springer-Verlag London Limited 2012

4.2 Mass-Spring-Damper Problem

Given a cube of mass $m = 1$ (kg) and of dimensions $2 \times 2 \times 2$ (m^3) (Fig. 4.1). The cube is connected to a spring of stiffness $K = 1$ (N/m) and a damping constant (friction) $C = 0.1$ (Ns/m).

The motion of the system is constrained in the horizontal plane of the ground. The cube oscillates along the x-direction (horizontal direction of the screen). Using Newton's second law, the following second-order differential equation for the x-displacement of the mass can be developed:

$$m\ddot{x}(t) + C\dot{x}(t) + Kx(t) = 0 \Rightarrow \ddot{x}(t) = -\frac{K}{m}x(t) - \frac{C}{m}\dot{x}(t) \qquad (4.1)$$

The purpose of this exercise is to animate the solution of this equation for $\dot{x}(0) = 2$ (m/s) and $x(0) = 0$ (m). Note that the mass is oscillating about the equilibrium position shown in Fig. 4.1 (nonextended position of the spring).

4.3 Creating the Virtual Scene for the Mass-Spring-Damper

The building blocks for this system are the mass, the spring, the wall (where the spring is attached), and the floor (Fig. 4.1). Since there are some shapes that are not available in the VRML, one might need to draw these shapes in more capable 3D software (such as 3ds Max® or AutoCAD®) and import these shapes into VRML. The spring shape was drawn in 3ds Max® and exported as a VRML file (the spring shape can be downloaded from Springer's web site http://extras.springer.com/ with the material of Chap. 4; the name of the file is *spring_shape.wrl*). The spring's length is 5. Before constructing the virtual scene, it is worth mentioning that all the objects (included in the library of VRML) will have their center positioned by default at the origin $(0,0,0)$. To create the virtual scene for the mass-spring-damper system, follow the following steps:

1. Open the downloaded file, *spring_shape.wrl*, using VRML.
2. To create the wall where the spring is attached, click on **Insert Box** (Fig. 4.2).
3. Click once on the title **Transform** to select it. Click another time on **Transform** and change the text to a meaningful name such as **Wall** (Fig. 4.3).
4. To change the size of the wall, double-click on the **Box** layer (Fig. 4.4). Then double-click the **size** layer. Tick the box of the x-axis, and then change the value to 0.1. Similarly, change the y- and z-axes values to 4. Click OK when done.
5. Click on the **New World** layer to select the top layer under which all the components should be positioned (Fig. 4.5).
6. Similar to step 2, click on **Insert Box** to create the mass.
7. Similar to step 3, change **Transform** to a meaningful name such as **Mass**.
8. Similar to step 2, click on **Insert Box** to create the floor and rename the **Transform** a meaningful name such as **Floor**.

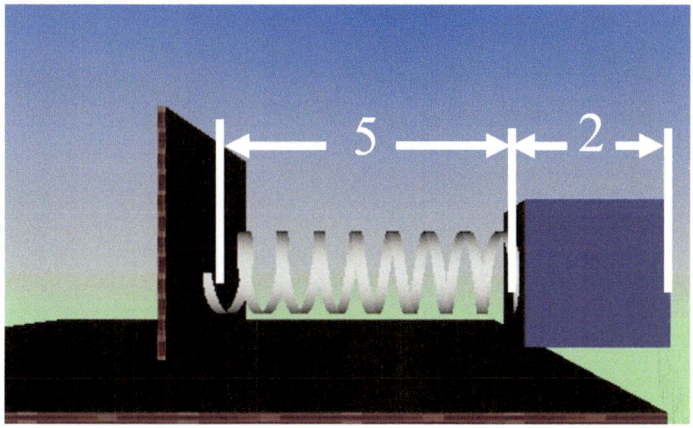

Fig. 4.1 Initial length of the spring in addition to the width of the mass

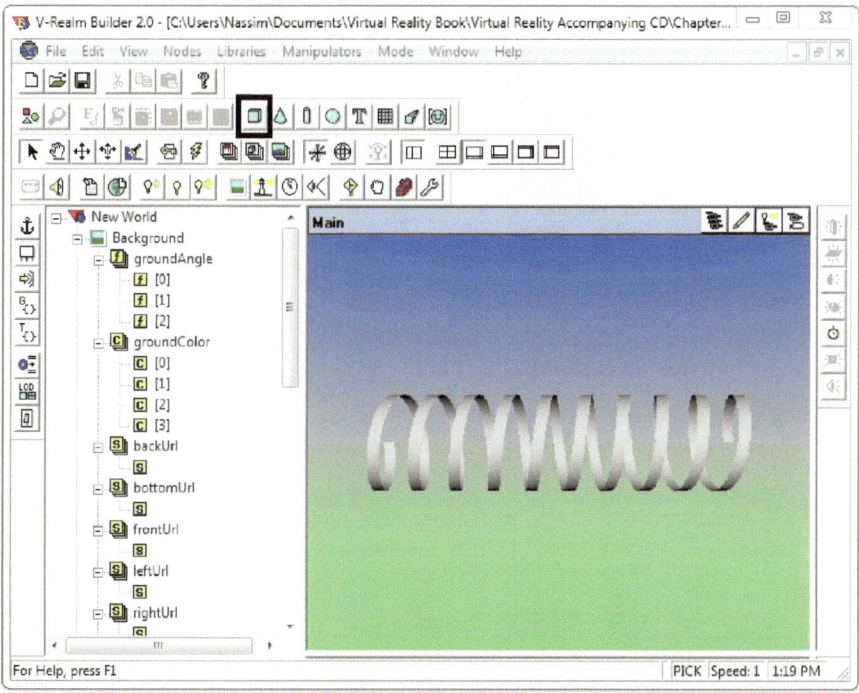

Fig. 4.2 Insert Box button is shown inside the rectangle

9. Repeat step 4 to resize the floor to 14, 0.1, and 10 in the *x*-, *y*- and *z*-directions, respectively. Click OK when done. The virtual scene should look like Fig. 4.6.

Figure 4.6 does not look like the final virtual scene. Elements such as the **Mass**, **Floor**, and **Wall** are centered by default at the point (0,0,0). The top surface of

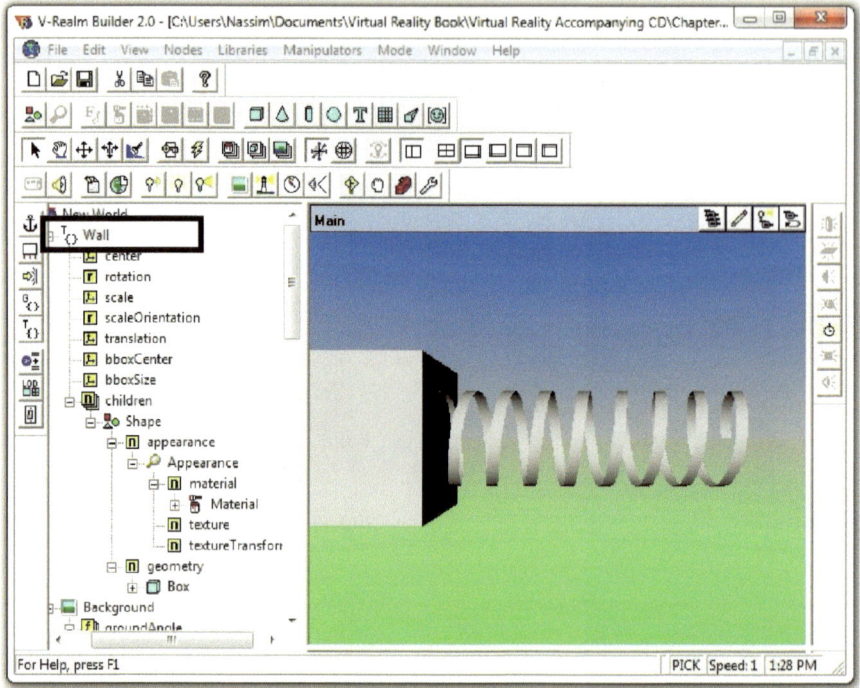

Fig. 4.3 Change the name of the cube from Transform to Wall

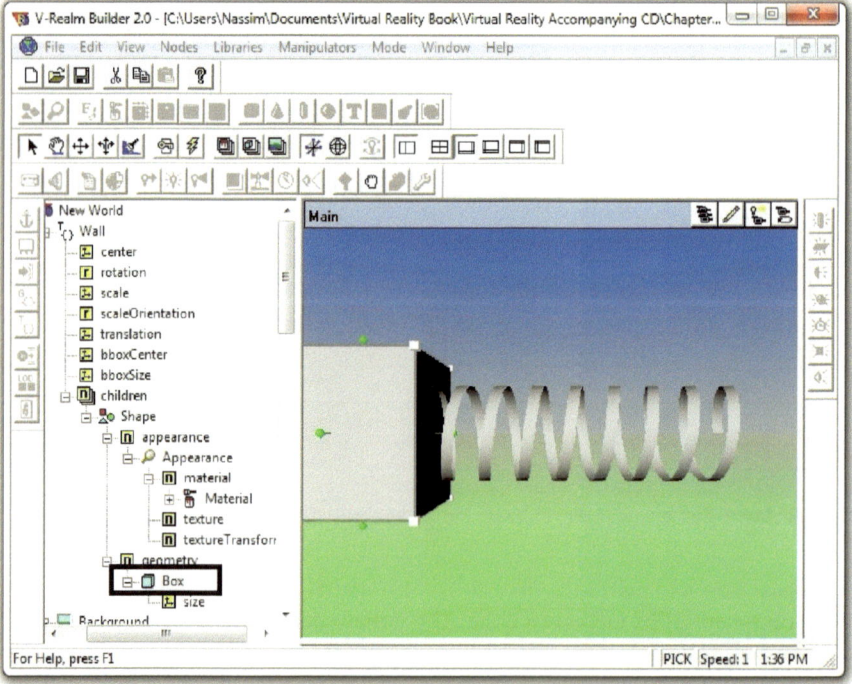

Fig. 4.4 Box is shown inside the rectangle

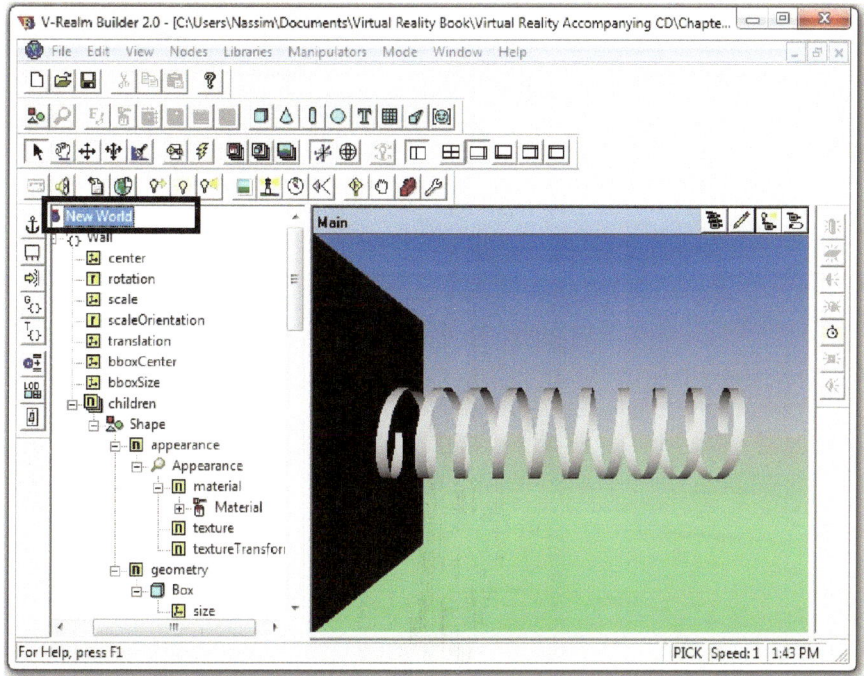

Fig. 4.5 New World layer

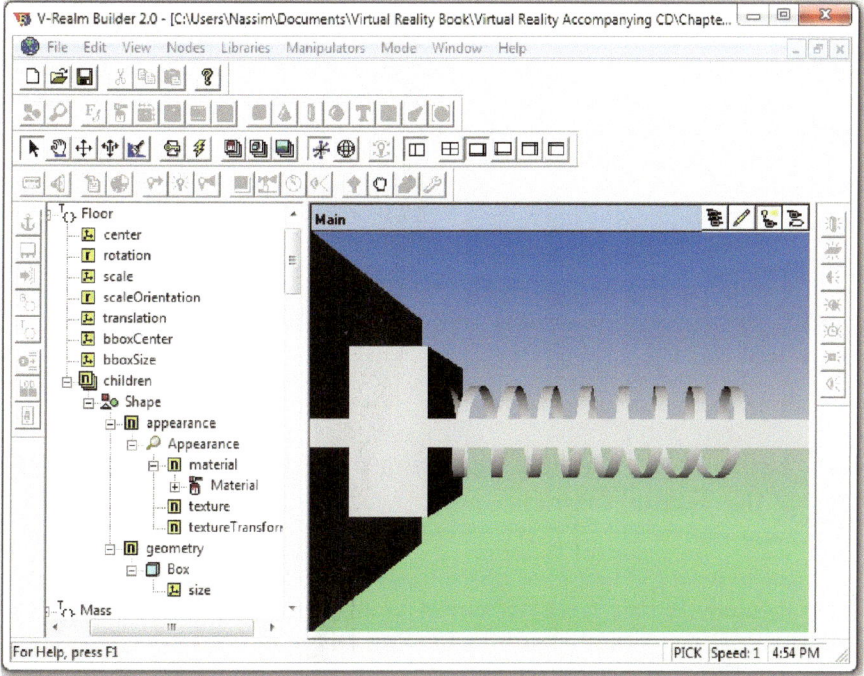

Fig. 4.6 Mass, Spring, Wall, and Floor

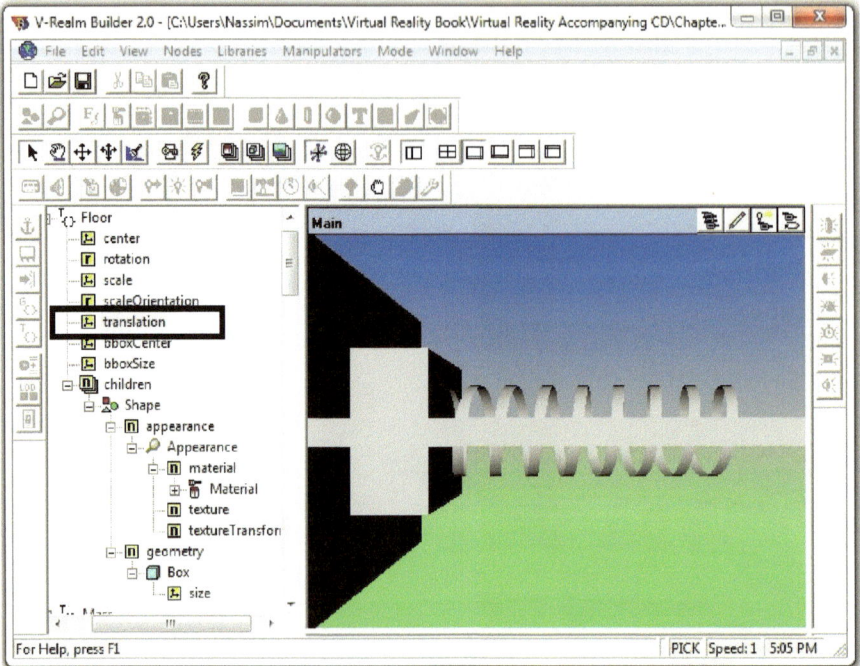

Fig. 4.7 The translation property is shown inside the rectangle

Floor should be touching the bottom surface of **Mass**. Thus, **Floor** should be translated in the *y*-direction (vertical direction) by half the height of the mass (the height of the mass is 2) plus half the thickness of the floor (which is 0.1). The overall translation is 1.05 downward. Double-click on **translation** (Fig. 4.7). Tick the box of the *y*-axis, and then change the value to −1.05. Click OK when done (Fig. 4.8). **Mass** should be translated to the other end of the spring, but since its position will be changed based on the solution of the differential Eq. 4.1, its translation will be handled by the M-script.

Click on **File > Save as**, and save it in the same directory where you will have your M-script. Name the file *virtual_scene*. The extension of the virtual reality file will be *.wrl*.

4.4 Adding Material and Changing the Color Properties of the Virtual Objects

To add a blue color to **Mass**, click on the **Color Mode** button (Fig. 4.9). The **Color Mode Painter** will show up. Double-click on the **Diffusive Color** area and then click to choose the desired color from the **Edit Diffusive Color** area (Fig. 4.10). Click OK when done. Notice that when you pass the cursor over the virtual scene,

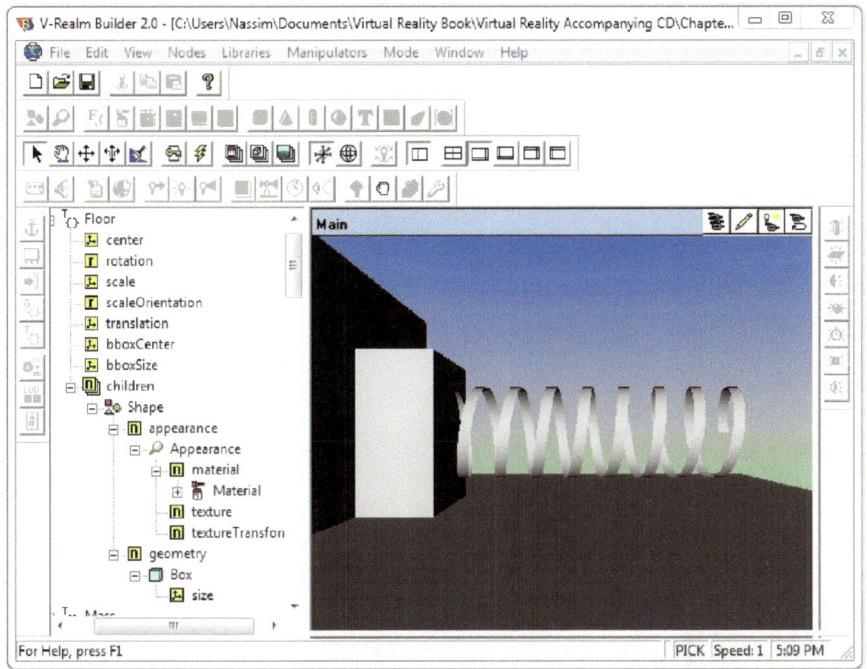

Fig. 4.8 Virtual scene with the translated Floor

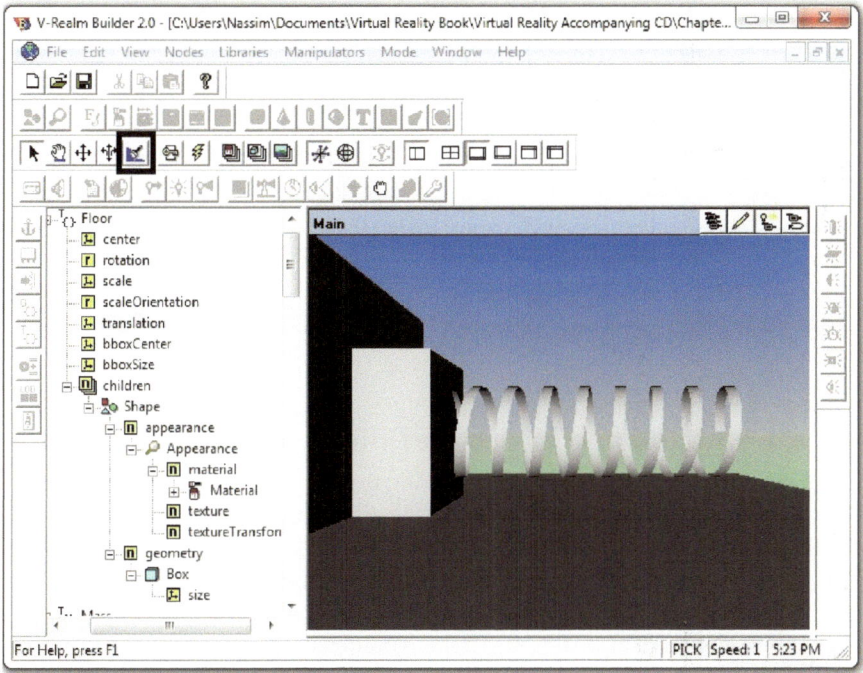

Fig. 4.9 Color Mode button is shown inside the rectangle

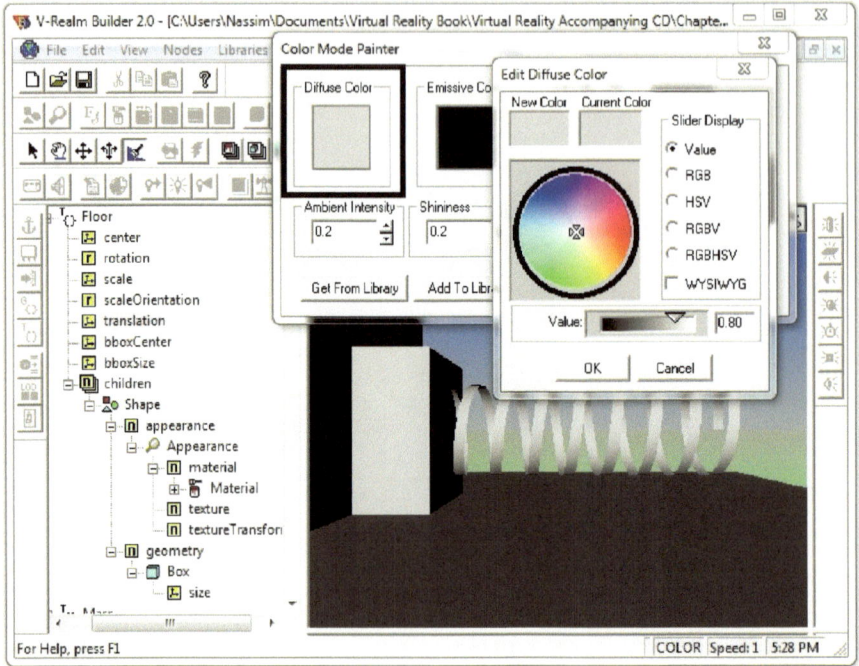

Fig. 4.10 Diffusive Color and Edit Diffusive Color areas in the rectangle and circle, respectively

it will change to a paint brush. Click on **Mass** in the virtual scene to color it. Close the **Color Mode Painter**.

To add a texture to **Floor** and **Wall**, click on the **Texture Library** button (Fig. 4.11). Choose the **Brick (Small)** texture (Fig. 4.12). Apply the texture on **Floor** by dragging the displayed texture inside the square to anywhere over **Floor** object in the virtual scene (click and hold the left mouse button to drag the texture, and then unclick it over the object to apply the texture). Similarly, apply the same texture on the wall. Click Close and save your file when done.

4.5   M-script File for Solving the Differential Equation

To numerically integrate for the solution of the differential equation given in Eq. 4.1, use the **ode45** function in MATLAB®. For more information about this function, type the following in the MATLAB® command window: **help ode45**.

The following M-script file, **MSD_solver.m,** will numerically integrate the differential equation given in Eq. 4.1. Some comments have been added to briefly explain the M-script file. The M-script can be found in Chap. 4 folder of the extra material of the book. The folder can be downloaded by following http://extras. springer.com/ and then searching for the book by entering the ISBN or the title.

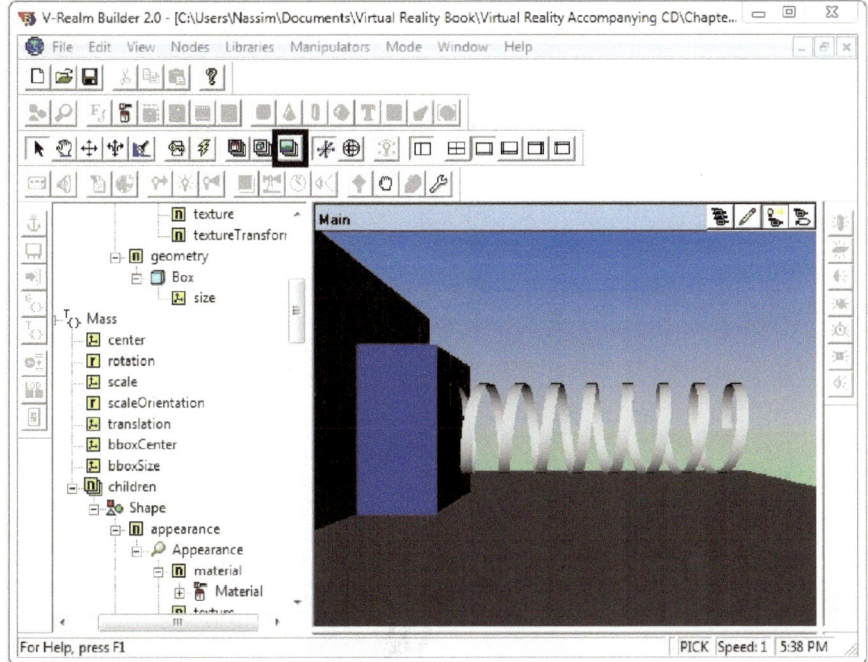

Fig. 4.11 Texture Library button is shown inside the rectangle

MSD_solver.m

```
%This file is the M-script differential equation solver
% for the mass_spring_damper example in Chapter 4:
t0=0;
%%%%%%%%%%%%%%%%%%%%%%%%%%%%%%%%%
% final time:
tf=100;
%%%%%%%%%%%%%%%%%%%%%%%%%%%%%%%%%
% Number of samples (for numerical integration of the solution)
N=1000;
%%%%%%%%%%%%%%%%%%%%%%%%%%%%%%%%%
% initial x position:
x_0=0;
%%%%%%%%%%%%%%%%%%%%%%%%%%%%%%%%%
% initial x velocity:
x_dot_0=2;
%%%%%%%%%%%%%%%%%%%%%%%%%%%%%%%%%
% Compute the solution of the differential equation
[t,x]=ode45(@myfunc,[t0:(tf-t0)/N: tf],[ x_0 x_dot_0]);
%%%%%%%%%%%%%%%%%%%%%%%%%%%%%%%%%
%plot the oscillations
plot(t,x(:,1));
```

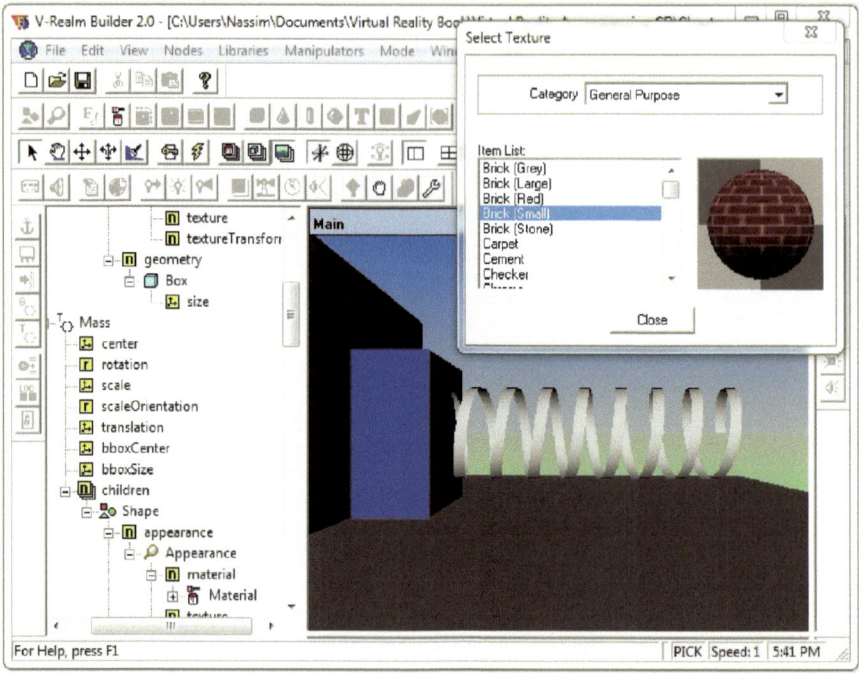

Fig. 4.12 Brick (Small) texture

The above M-script file, ***MSD_solver.m***, calls ***myfunc.m*** which computes the derivative vector $[x(t)\,\dot{x}(t)]$. The following is the script for ***myfunc.m***:

myfunc.m

```
function dxdt=myfunc(t,x)
%%mass
m=1;
%%daming coefficient
C=0.1;
%%spring stiffness constant
K=1;
%% states:
%x_dot
dxdt_1=x(2);
%x_dot_dot
dxdt_2=-K/m*x(1)-C/m*x(2);
%derivative vector
dxdt=[dxdt_1;dxdt_2];
```

4.6   M-script File for Animating the Virtual Scene

In what follows, the virtual scene, *virtual_scene.wrl*, will be animated based on the solution given by *MSD_solver.m*, and the animation will be recorded into a movie file, *MSD.avi* (the reader should run *MSD_solver.m* before running *MSD_animate.m*). The parameters that will be changed in the virtual world are the **translation** property of the mass and the length of the spring. The default position of the center of the mass point is at $(0,0,0)$. The instantaneous x-position of the center of the mass is equal to the original (unextended) length of the spring plus half the length of the mass plus the oscillation of the mass around its equilibrium position given by Eq. 4.1. Thus, the instantaneous x-position of the mass is equal to $5 + 2/2 + x(t)$. Note that the wall is positioned at point $(0,0,0)$. As for the length of the spring, it will be changed by scaling it in the x-direction according to the instantaneous position of the mass.

A first look at *MSD_animate.m* script might seem overwhelming and full of commands; therefore, the reader is advised to start writing his/her own code by modifying an existing one instead of starting from scratch every time.

The script that comes before the for-loop consists of three parts. The first (refer to Part 1 of *MSD_animate.m*) one deals with opening and viewing the virtual world. The second (refer to Part 2 of *MSD_animate.m*) deals with setting the recording of the animation to a *.avi file. In this example, the movie will be saved as *MSD.avi*. As for the third part (refer to Part 3 of *MSD_animate.m*), it extracts the original scale of the spring in x-, y-, and z-directions and stores the values in scale_x, scale_y, and scale_z variables. Originally, the spring has the x-, y-, and z-scales equal to 1.

As for the for-loop, the first parameter that will be changed in the virtual world is the **translation** property of the mass. Six is equal to the original length of the spring plus half the length of the mass in the x-direction (Fig. 4.1). x(Index,1) is the elongation of the spring $x(t)$ given by Eq. 4.1:

world.Mass.translation=[6+x(Index,1) 0 0];

To vary the length of the spring, its scale in the x-direction should be changed according to the instantaneous position of the mass. The scale in the x-direction is given by

$$scale\_x = (5 + x(t))/5 \tag{4.2}$$

Equation 4.2 is implemented in the for-loop by writing:

world.Spring.scale=[(5+x(Index,1))/5 scale_y scale_z];

After the reader runs *MSD_animate.m,* a video file *MSD.avi* should be created in the same working directory. The reader should try playing the movie to see a playback of the animation. To increase or decrease the playtime of the movie, the reader can change the number of samples used for the numerical integration of the differential Eq. 4.1.

MSD_animate.m

```
%%%% Part 1
% creates a virtual world associated with virtual_scene.wrl file
world = vrworld('virtual_scene');
%open the virtual world
open(world);
%draw the virtual world
fig = view(world, '-internal');
%
%
%
%%%% Part 2
%defining the interval time to record the animation
set(world, 'RecordInterval', [0 tf]);
%defining the name of the movie file to be cube.avi
set(fig, 'Record2DFileName','MSD.avi');
% enabling 2D recording
set(fig, 'Record2D', 'on');
%compression quality of the movie
set(fig, 'Record2DCompressQuality', 100);
%defining recording mode to be scheduled
set(world, 'RecordMode', 'scheduled');
%
%
%
%%%% Part 3
%computing the original scale of the spring
Scale=world.Spring.scale;
scale_x=Scale(1);
scale_y=Scale(2);
scale_z=Scale(3);

%initiating the counter

Index=0;
%sweeping the time interval
for t=t0:(tf-t0)/N:tf
    %incrementing the counter
    Index=Index+1;
    %changing the translation property of the object Cube that was
    %defined in virtual_scene.wrl to the value computed by ode45
    world.Mass.translation=[6+x(Index,1) 0 0];
    %computing the scale factor=(original length+delta_x)/original
    %length
    world.Spring.scale=[(5+x(Index,1))/5 scale_y scale_z];
    %changing the time of the virtual world to match the time that was
    %given by ode45
    set(world,'Time', t);
    %update the virtual reality scene
    vrdrawnow;
end
%close the virtual world
close(world);
delete(world);
```

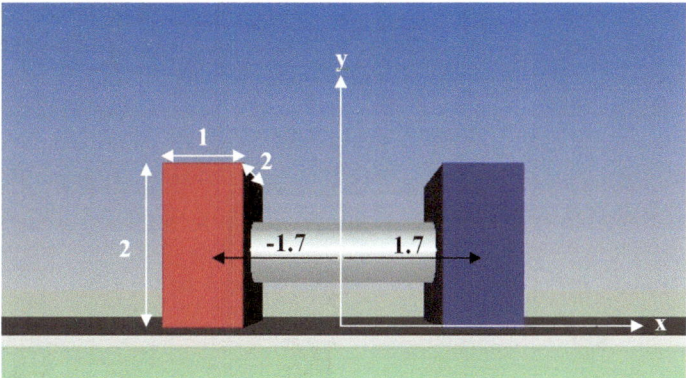

Fig. 4.13 Dimensions of the two masses and their initial positions

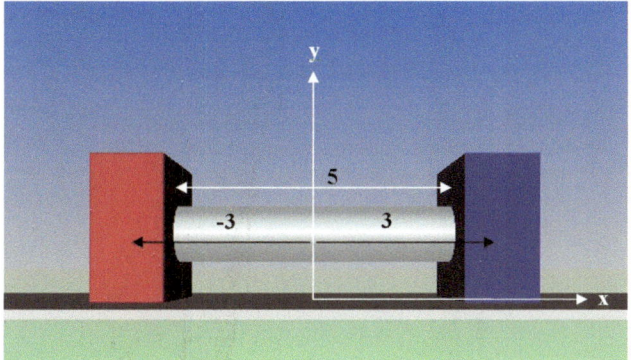

Fig. 4.14 Unextended position of the spring with the two masses

4.7   Application Problem

Given two masses $m_{B1} = 1\,(\text{kg})$ and $m_{B2} = 1\,(\text{kg})$ of the same dimensions $1 \times 2 \times 2\,(\text{m}^3)$ (Fig. 4.13). The two masses are connected by a spring of stiffness $K = 1\,(\text{N/m})$ and a damping constant (friction) $C = 0.15\,(\text{Ns/m})$. The centers of m_{B1} and m_{B2} are initially situated at $x_{B1}(0) = -1.7\,(\text{m})$ and $x_{B2}(0) = 1.7\,(\text{m})$, respectively (Fig. 4.13). Both masses have zero initial velocities. The initial unextended length of the spring is $5\,(\text{m})$ (Fig. 4.14). Assuming no friction with the ground, solve the following:

1. Reconstruct the virtual scene shown in Fig. 4.13 in VRML assuming that the bottom faces of the masses are located at $y = 0$ and that the two boxes are aligned horizontally. Add an additional box component to represent the ground where the top face of this box is at $y = 0$. Represent the spring by a cylinder object or initial radius $r_0 = 0.4\,(\text{m})$.

2. Develop the equations of motion of the two masses (Hibbler 1986).
3. Develop the M-script file to compute the numerical solution of the equations of motion, animate the VRML model, and record the simulation into a video. The objects that should be animated in the scene are the two masses and the spring. The spring should be scaled in the x-direction such that its initial volume does not change. Assume the total time of simulation is 50 s and that the step size is 0.01 s.

The solution of the problem can be downloaded from Springer's web site http:// extras.springer.com/. The files can be found in folder /Chapter 4/Application Problem. Run the script *Two_masses_solver.m* to generate the solution.

Reference

Book

Hibbler RC (1986) Engineering mechanics: dynamics, 4th edn. McMillan, New York

Chapter 5
Crank-Slider Mechanism of a Piston

5.1 Introduction

The purpose of this chapter is to give MATLAB® users a better understanding on how to animate a physical system that has more than one component in a virtual reality environment. This chapter will also help the reader to implement a simple PID controller to control a ball on a plate and visualize the performance of the controller.

In Sect. 5.2, a crank-slider mechanism problem will be presented.

In Sect. 5.3, step-by-step instructions will guide the reader through drawing the virtual scene in VRML. The components of the system that will be drawn are the crank, the connecting rod, the piston, and the cylinder.

Section 5.4 will include a description of the M-script file that will animate the virtual scene. In addition, the script will record the animation into a video file.

The chapter concludes by a control problem for a ball on a plate. The position of the ball is controlled on the plate by means of two independent actuators. The problem will include developing the equations of motion of the ball, constructing the virtual scene, developing the two PID controllers for the plate, and animating the ball and the plate.

The electronic version of all the M-script files and VRML models in addition to the recorded movies for the animated crank-slider mechanism and the controlled motion of the ball on the plate can be downloaded from Springer's web site http://extras.springer.com/.

5.2 Crank-Slider Mechanism

Figure 5.1 shows a sketch for the crank-slider mechanism. Point O is the center of the crankshaft. ON is the crankshaft. Its length is $r = 10$ (cm). NP is the connecting rod. Its length is $l = 20$ (cm). Since P is constrained to move in the vertical direc-

---

Matlab® and Simulink® are registered trademarks of The Mathworks, Inc.

N. Khaled, *Virtual Reality and Animation for MATLAB® and Simulink® Users:* 35
Visualization of Dynamic Models and Control Simulations,
DOI 10.1007/978-1-4471-2330-9_5, © Springer-Verlag London Limited 2012

Fig. 5.1 A sketch for the
crank-slider mechanism

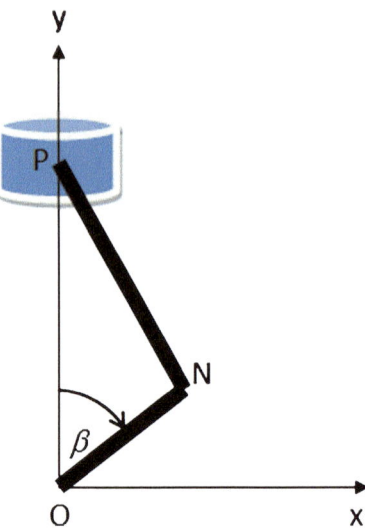

tion and the rest of the components are coplanar, this system has one degree of
freedom which will be considered to be the crankshaft angle β.

The coordinates for point O, which is the pivot point for the crankshaft, are
$(0,0,0)$. The coordinates of the point N are $(r\sin(\beta), r\cos(\beta), 0)$. The coordinates
of the point P are $(0, r\cos(\beta) + \sqrt{l^2 - r^2\sin^2(\beta)}, 0)$.

5.3 Creating the Virtual Scene for the Crank-Slider Mechanism

The building blocks for this system are the crank, the rod, the piston, and the cylinder.
To create the virtual scene for the crank-slider mechanism, follow the following steps:

1. Open the V-Realm Builder and click on **File>New** to create a new virtual
   world.
2. Click on the **Insert Background** button (Fig. 5.2) to create a background for
   the scene.
3. Click on the **Insert Cylinder** (Fig. 5.3) to create the crankshaft. In the main
   view window, you should see a cylinder with a green and blue background
   behind it. Notice that on the left side of the main view window, some text and
   yellow icons have appeared, all under the title **Transform**.
4. Click once on the title **Transform** to select it. Click another time on **Transform**
   and change the text to a meaningful name such as **Crank** (Fig. 5.4).
5. To change the dimensions of the crankshaft, double-click on **Cylinder** icon to
   expand its corresponding sub-icons (Fig. 5.5). Then double-click the height property
   of the cylinder and change it to 1. Similarly, change the radius to 0.2 (in Fig. 5.5,
   under the **Cylinder** icon, you will have the geometrical properties of the cylinder).

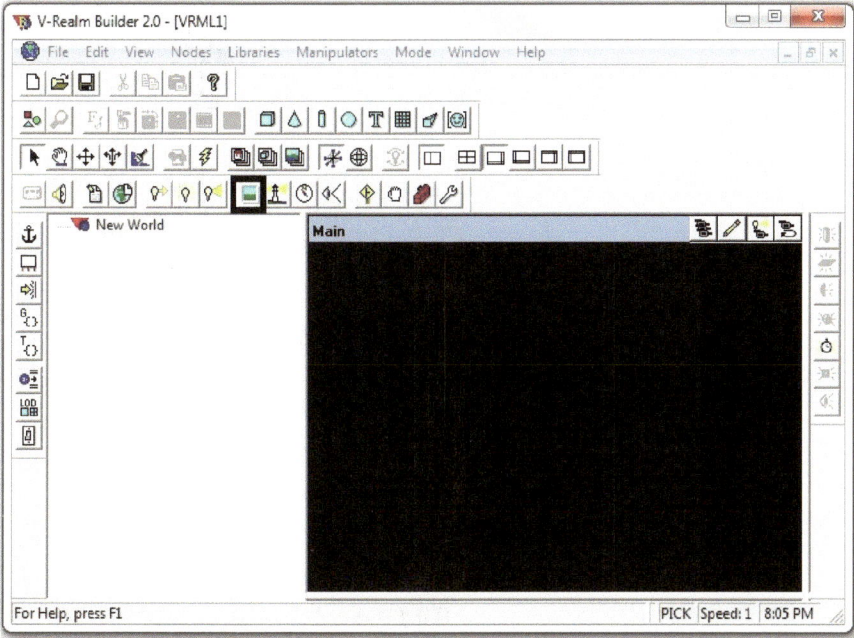

Fig. 5.2 Insert Background button is shown inside the rectangle

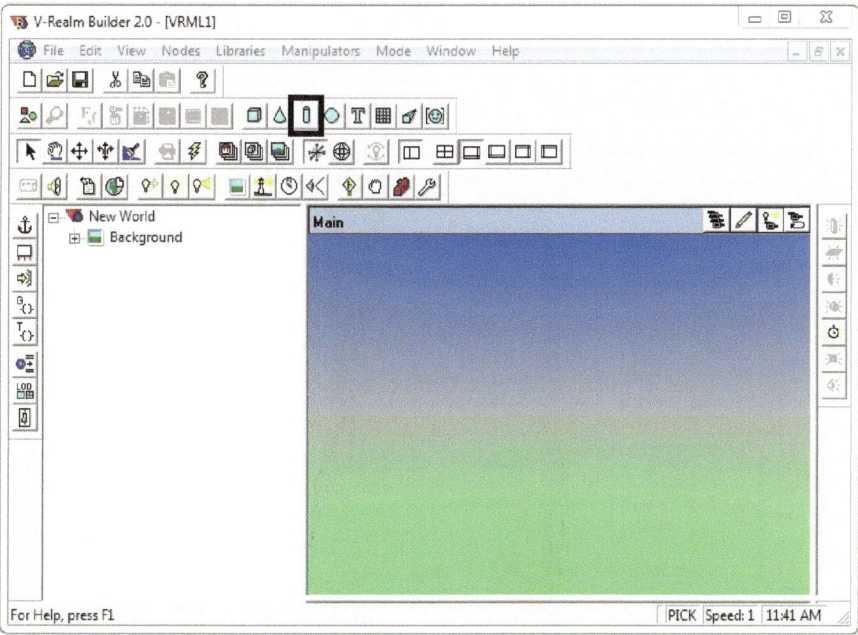

Fig. 5.3 Insert Cylinder button is shown inside the rectangle

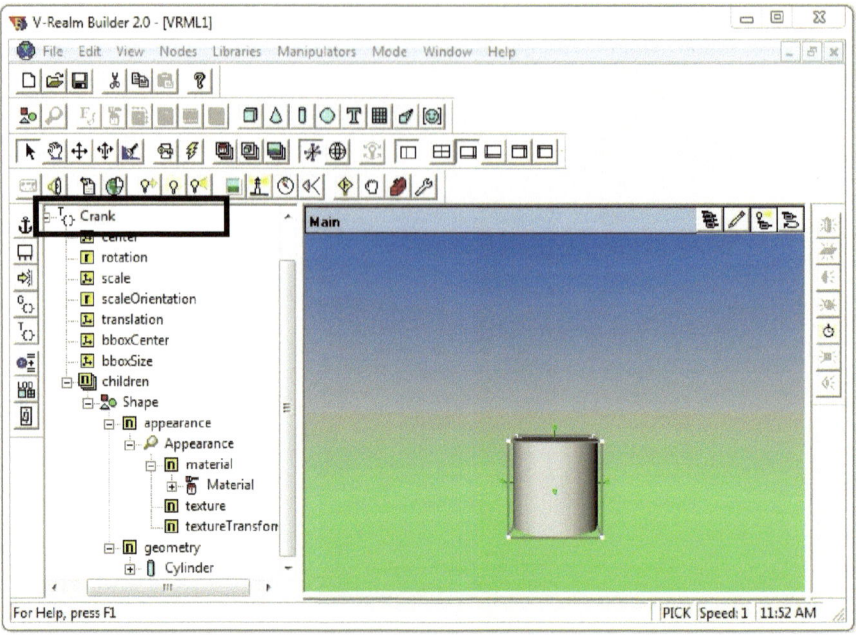

Fig. 5.4 Change the name of the cylinder from Transform to Crank

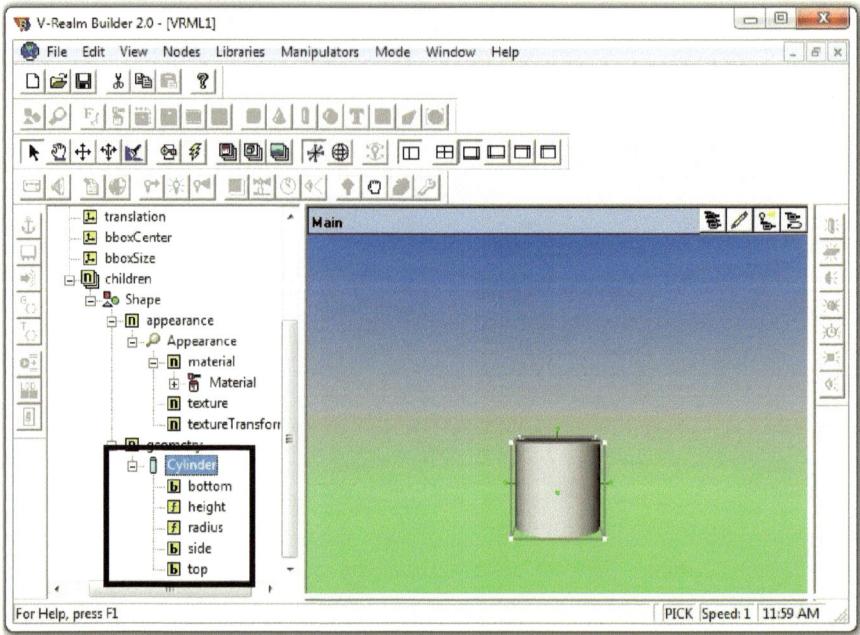

Fig. 5.5 The height and radius properties are shown inside the rectangle (under Cylinder)

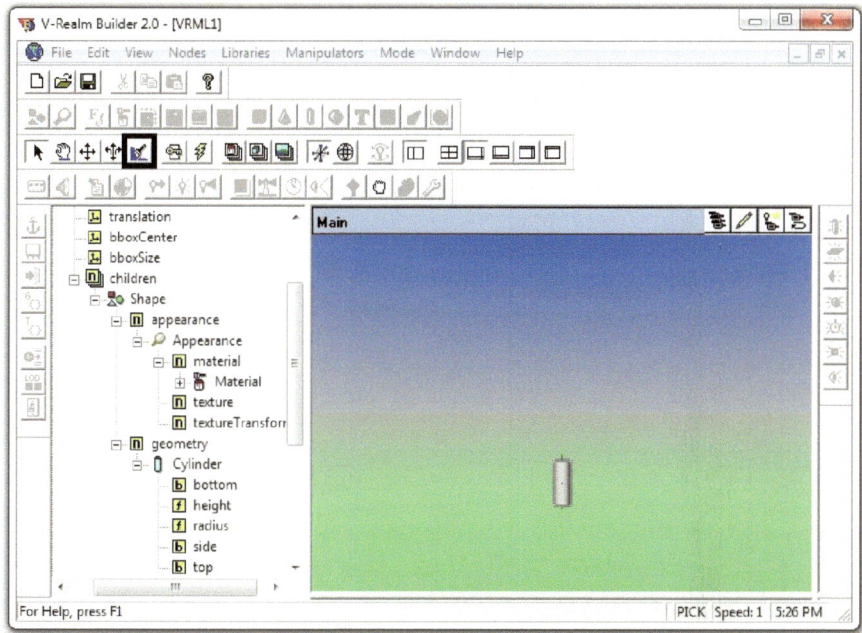

Fig. 5.6 Color Mode button is shown inside the rectangle

6. Click on the **Color Mode** button (Fig. 5.6). The **Color Mode Painter** will show up. Double-click on the **Diffusive Color** area and then click to choose the desired color from the **Edit Diffusive Color** area (Fig. 5.7). Click OK when done. Notice that when you pass the cursor over the virtual scene, it will change to a paint brush. Click on **Crank** in the virtual scene to color it. Close the **Color Mode Painter**.
7. Click on the **New World** layer to select the top layer under which all the components should be added (Fig. 5.8).
8. Click on the **Insert Cylinder** (Fig. 5.3) to create the connecting rod. Rename the created **Transform Rod**.
9. Repeat step 5 to change the height property of **Rod** to 2 and the radius to 0.2.
10. Repeat step 7 to select the top layer, **New World**.
11. Click on the **Insert Cylinder** (Fig. 5.3) to create the piston. Rename the created **Transform Piston**.
12. Repeat step 5 to change the height property of **Piston** to 0.5 and the radius to 0.5.
13. Click on the **New World** layer to select the top layer under which all the components should be positioned (Fig. 5.8).
14. Click on the **Insert Cylinder** (Fig. 5.3) to create the cylinder. Rename the created **Transform Cylinder**.
15. Repeat step 5 to change the height property of the cylinder to 1.2 and the radius to 0.5.
16. To make **Cylinder** semitransparent, click on the+sign on the left of **Material** icon to expand its corresponding sub-icons (Fig. 5.9). Then double-click the **transparency** property (Fig. 5.10) and change it to 0.5. Click OK when done.

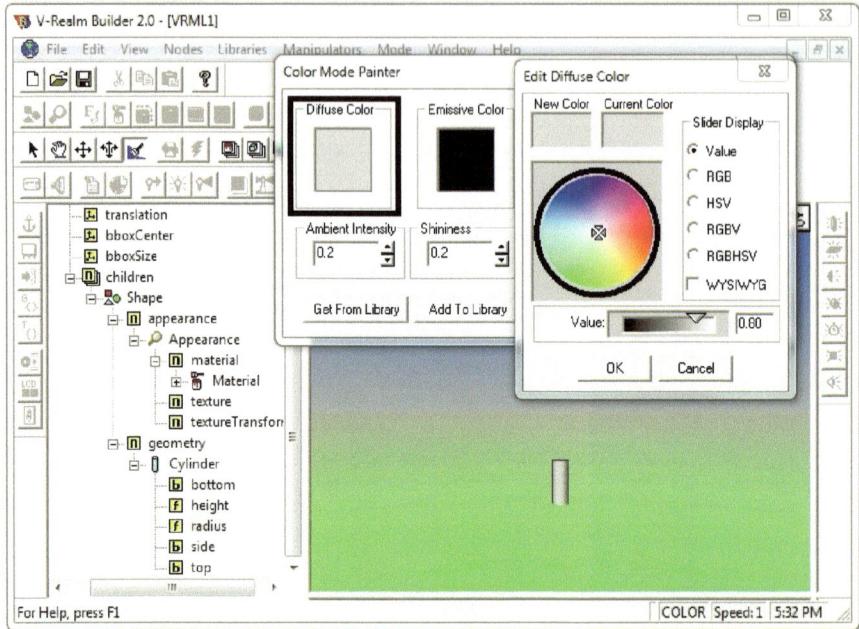

Fig. 5.7   Diffusive Color and Edit Diffusive Color areas in the rectangle and circle, respectively

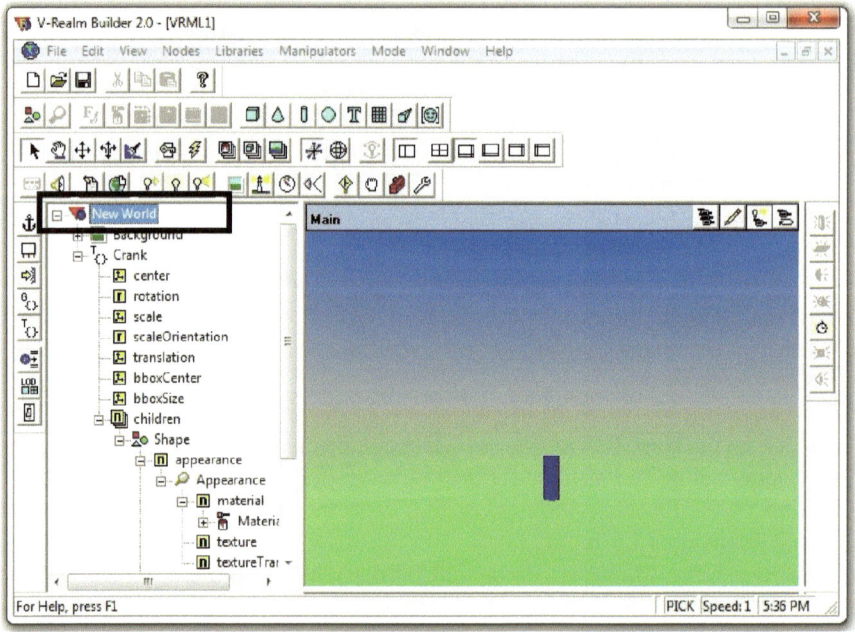

Fig. 5.8   New World layer

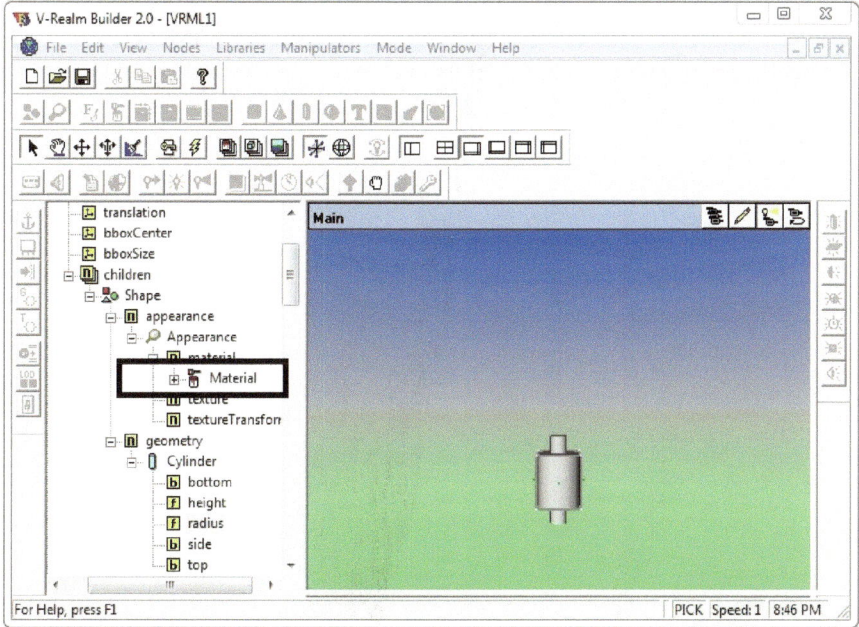

Fig. 5.9 Material is shown inside the rectangle

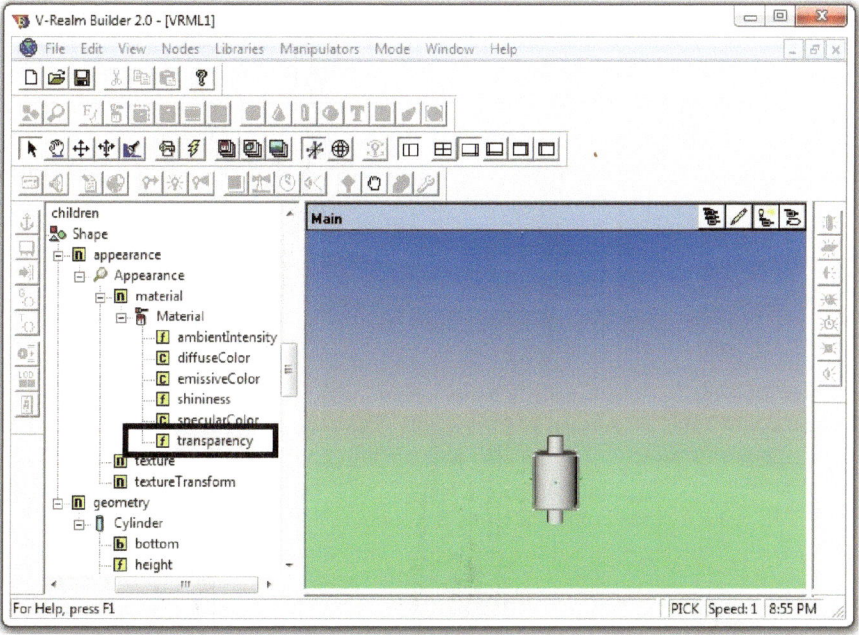

Fig. 5.10 The transparency is shown inside the rectangle

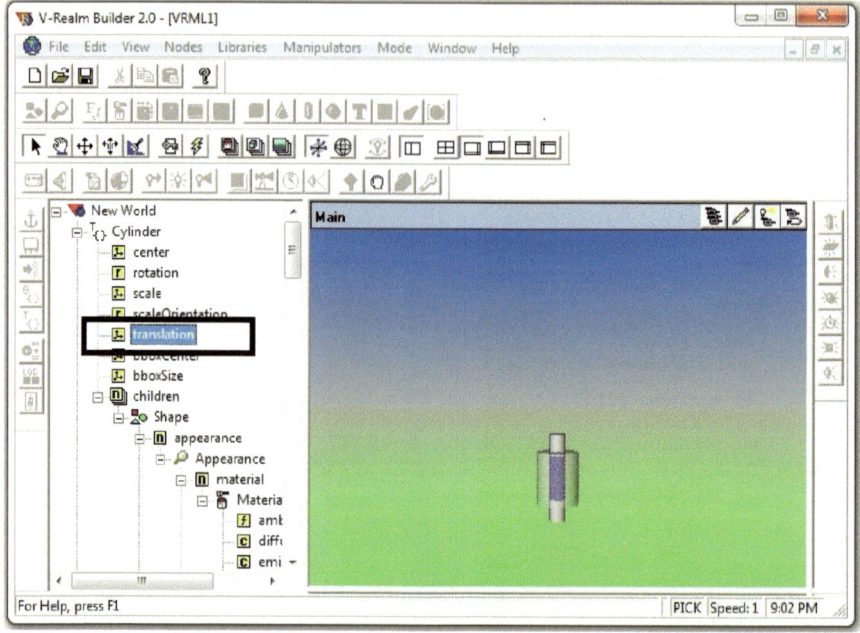

Fig. 5.11 The translation property is shown inside the rectangle

17. Translate **Cylinder** upward by double-clicking on **translation** property
    (Fig. 5.11), then ticking the square beside the y-axis value and changing the
    value to 3. Click OK when done.
18. Click on **File>Save as** and save it in the same directory where you will have
    your M-script files. Name the file *virtual_scene*. The extension of your Virtual
    Reality file will be *.wrl*.

5.4 M-script File for Animating the Virtual Scene

In what follows, the virtual scene, *virtual_scene.wrl*, will be animated by assuming
that the crank angle is an input to the system. The animation will be recorded into a
movie file *crank_slider.avi*. Figure 5.12 shows the centers (midpoints) M_1 and M_2 of
the crankshaft and the connecting rod, respectively. In addition, Fig. 5.12 shows β
and γ, the angles of the crankshaft, and the connecting rod with the vertical, respectively. The parameters that will be changed in the virtual world are the following:

1. The **translation** property of **Piston**
2. The **rotation** property of the **Rod** (around its center M_2)
3. The **translation** property of **Rod** (translation of its center M_2)

Fig. 5.12 M_1 and M_2 are the centers (midpoints) of the crank and connecting rod, respectively

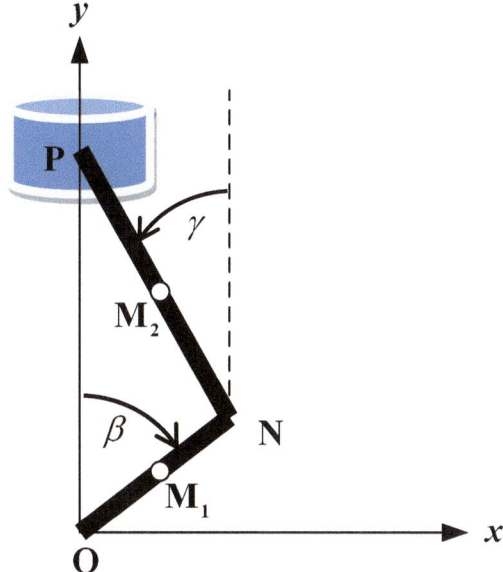

4. The **rotation** property of **Crank** (around its center M_1)
5. The **translation** property of **Crank** (translation of its center M_1)

One might ask: How to define the centers M_1 and M_2 of **Crank** and **Rod**? When the virtual scene was created, the crank in VRML, by default the center of rotation, is placed at M_1 which is the midpoint of ON. The **bboxCenter** property (Fig. 5.13) defines the center of rotation of the objects, and by default it is set to $(0,0,0)$ at the center of the geometrical object. In this particular example, the centers M_1 and M_2 will be left at their default position (at the midpoint of the cylinders).

The script that comes before the for-loop consists of two parts. The first (refer to Part 1 of *crank_slider.m*) is the one dealing with opening and viewing the virtual world. The second (refer to Part 2 of *crank_slider.m*) deals with setting the recording of the animation to a **.avi* file. In this example, the movie will be saved as *crank_slider.avi*.

In the for-loop, the crank angle, β, (Fig. 5.12) is being varied from 0 to 4π rad with increments of 0.08 rad. The rod angle, γ, (Fig. 5.12) is being computed accordingly. Note that both angles are measured with respect to the vertical line (a zero angle means that the object is aligned with the vertical line). The translation properties of the piston, the crankshaft, and the rod are varied. The rotation properties of the crankshaft and the rod are also being varied. Refer to Sect. 5.2 to see how the coordinates of points N and P are being computed.

After you run *crank_slider.m*, a video file *crank_slider.avi* should be created in the same directory as the MATLAB file. Try playing the movie to see a playback of the animation. To increase or decrease the playtime of the movie, change the crank angle increment in the for-loop.

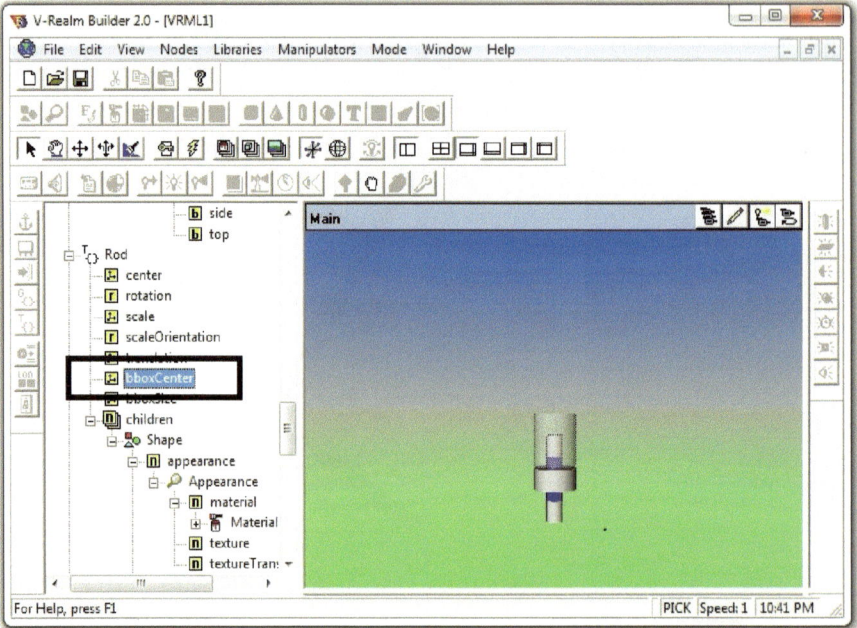

Fig. 5.13 The bboxCenter property defines the center of the object

crank_slider.m

```
%%% Part 1
% creates a virtual world associated with virtual_scene.wrl file
world = vrworld('virtual_scene');
%open the virtual world
open(world);
%draw the virtual world
fig = view(world, '-internal');
%
%
%
%%% Part 2
%defining the interval time(which is based on the angle of the
%  crank shaft) to record the animation
set(world, 'RecordInterval', [0 4*pi]);
%defining the name of the movie file to be cube.avi
set(fig, 'Record2DFileName','crank_slider.avi');
% enabling 2D recording
set(fig, 'Record2D', 'on');
%compression quality of the movie
```

```
set(fig, 'Record2DCompressQuality', 100);
%defining recording mode to be scheduled
set(world, 'RecordMode', 'scheduled');


%The crank-slider mechanism sketch

%   P
%   o
%    o
%     o
%      o N
%      *
%      *
%      *
%   O

% O [0,0,0] is the pivot of the crank shaft

%defining the geometrical properties l and r
l=2;
r=1;
%initiating the counter that will be used in the for loop
i=0;

for A=0:0.08:4*pi
   i=i+1;
   y=r*cos(A)+sqrt(l^2-r^2*sin(A)^2);
   %define the coordinates of point N
   N=[r*sin(A) r*cos(A) 0];
   %define the coordinates of point P
   P=[0 y 0];
   %angle of the rod with the vertical y-axis
   rod_angle=atan2((P(1)-N(1)),(P(2)-N(2)));
   %crank angle is being varied
   crank_angle=A;
   %translating the piston
   world.Piston.translation=P;
   %translating crank
   world.Crank.translation=N/2;
   %rotating the crank (- sign for clockwise rotation)
```

```
        world.Crank.rotation=[0 0 1 -crank_angle];
        %translating the rod
        world.Rod.translation=(N+P)/2;
        %rotating the rod(- sign for clockwise rotation)
        world.Rod.rotation=[0 0 1 -rod_angle];
        set(world,'Time', A);
        vrdrawnow;
    end
    %close the virtual world
    close(world);
    delete(world);
```

5.5   Application Problem: Control of a Ball on a Plate

Given a ball of mass $m = 0.008$ (kg) and radius $r_0 = 0.5$ (m) (Fig. 5.14). The plate is of dimensions $7 \times 0.2 \times 7$ (m³). The plate is pivoted around the origin $O(0,0,0)$ and has two independent actuators that can rotate it around the x and z global axes by angles β and α, respectively (-0.5 rad $\leq \beta \leq 0.5$ rad and -0.5 rad $\leq \alpha \leq 0.5$ rad). Assuming no friction and no slipping between the plate and the ball, solve the following:

1. Develop the two equations of motion of the two degrees of freedom ball (Hibbler 1986). Where I_0 is the moment of inertia of the ball, prove that the equations of motions are

$$\left(m + \frac{I_0}{r_0^2} \right) \ddot{x} - m \left(\dot{\alpha}\dot{\beta}z + \dot{\alpha}^2 x \right) + mg \sin(\alpha) = 0, \tag{5.1}$$

$$\left(m + \frac{I_0}{r_0^2} \right) \ddot{z} - m \left(\dot{\alpha}\dot{\beta}z + \dot{\beta}^2 x \right) + mg \sin(\beta) = 0. \tag{5.2}$$

2. Neglecting the actuator dynamics (i.e., assuming a commanded angle α or β can be instantly delivered by the actuators), and assuming no coupling between the two actuators, develop the two PID controllers (Ogata 2010) for the angles α and β of the plate to bring the ball to origin within a settling time of 50 and 70 s for x and z, respectively. (Hint: define the error for α and β controllers to be $(x-0)$ and $(z-0)$, respectively.) Write the M-script file to simulate the two controllers along with numerical integration of Eqs. 5.1 and 5.2. Set the total time of simulation to 80 s and the step size to 0.05 s. The initial conditions for system are:

$$x(0) = 1.3 \ (m)$$

Fig. 5.14 Ball on a plate in
addition to the rotation angles
of the plate

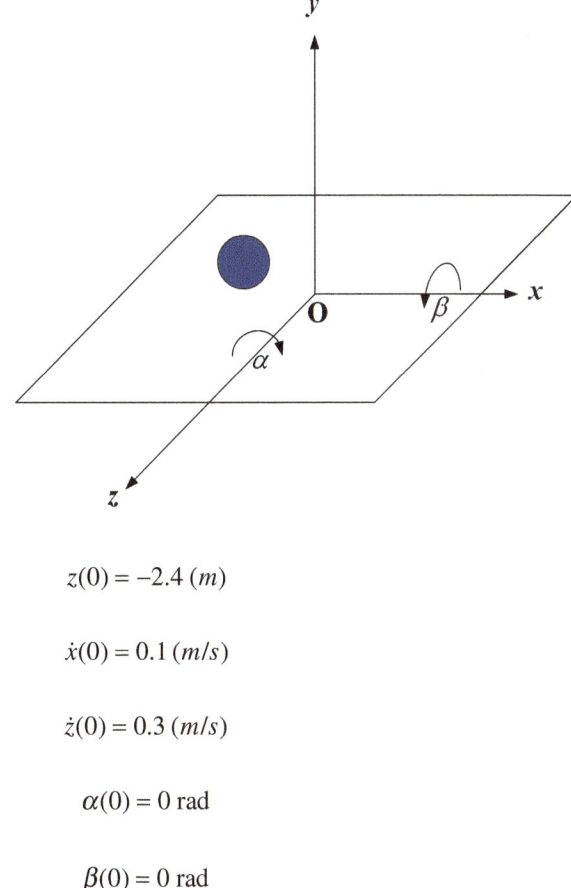

$$z(0) = -2.4 \ (m)$$

$$\dot{x}(0) = 0.1 \ (m/s)$$

$$\dot{z}(0) = 0.3 \ (m/s)$$

$$\alpha(0) = 0 \ \text{rad}$$

$$\beta(0) = 0 \ \text{rad}$$

3. Plot x and z versus time on the same plot. Also plot the angles α and β versus time on the same plot.
4. Reconstruct the virtual scene shown in Fig. 5.15 in VRML.
5. Write an M-script file that relies on the time vectors (x, z, α, and β) generated in Part 2 and the VRML model constructed in Part 4 to animate the ball and plate. The following are some tips to help the reader develop the code:

   - The reader has to construct the rotation matrix by using the two rotation angles α and β.
   - The reader has to watch out for the sign convention for the rotation angles (α and β) when constructing the rotation matrix. From Fig. 5.14, a positive rotation β of the plate around x-axis would ultimately yield a positive z-displacement of the ball. Similarly, a positive rotation α around the z-axis would ultimately yield a positive x-displacement of the ball. β is oriented anticlockwise for its positive direction, while α is oriented clockwise for its positive direction.

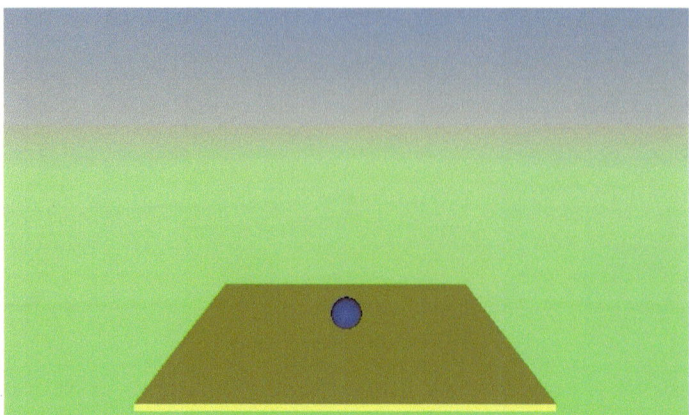

Fig. 5.15 Virtual scene of the ball and the plate

- Once the rotation matrix has been constructed, the reader is advised to use the function **vrrotmat2vec.m** to change the rotation matrix (3×3) to a rotation vector (4×1). The rotation vector can be used as an input to the VRML model to rotate the plate.
- The reader has to compute the global y-position of the ball. To do so, the reader can use the normal vector of the plate in addition to the fact that the center of the ball is always at a perpendicular distance equal to the radius of the ball. Once y is computed, the global position of the ball in space can be changed in VRML by changing the translation property of the ball to (x, y, z).

The solution of the problem can be downloaded from Springer's web site http://extras.springer.com/. The files can be found in folder/Chapter 5/Application Problem.

References

Books

Hibbler RC (1986) Engineering mechanics: dynamics, 4th edn. McMillan, New York
Ogata K (2010) Modern control engineering, 5th edn. Prentice Hall, New York

Chapter 6
Car Animation with Joystick Control

6.1 Introduction

This chapter aims to teach the user how to handle a set of 3D rotations and translations of VRML bodies. In addition, the reader will get to know how to create and handle various viewpoints in VRML. Furthermore, the user will learn how to utilize human interface devices (such as the joystick) to control and interact with the VRML bodies.

There are many ways to describe 3D rotations of bodies. For example:

- Euler rotations (represented by three angular rotations)
- Rotation vector (represented by a unit vector, u, and a rotation angle, ϕ)

An Euler rotation around the x_1-axis by an angle ϕ is identical to the rotation vector $\begin{bmatrix} 1 & 0 & 0 & \phi \end{bmatrix}$. This fact will be used to model a series of complex 3D rotations by representing the rotation vector as cascaded Euler rotations. In VRML environment, rotations are represented using the rotation vector where the rotation angle is expressed in degrees (when passing angles from MATLAB® and Simulink® to VRML, angles should be in radians).

In this chapter, the motion of a car in the horizontal plane will be animated. The steering of the car and the traction force will be controlled by means of a joystick.

In Sect. 6.2, the governing equations of motion of the car in the horizontal plane will be introduced. The car is subject to a steering action and traction force. The various degrees of freedom of the system will be defined along with the discrete empirical equations of motion.

In Sect. 6.3, a step-by-step guide to draw the virtual scene for the car and a straight road in VRML will be presented. In addition, two **Viewpoints** (points where the viewer is located at the scene) will be created in VRML.

Section 6.4 will guide the reader to set up the joystick that will be used to control the car.

Matlab® and Simulink® are registered trademarks of The Mathworks, Inc.

N. Khaled, *Virtual Reality and Animation for MATLAB® and Simulink® Users:*
Visualization of Dynamic Models and Control Simulations,
DOI 10.1007/978-1-4471-2330-9_6, © Springer-Verlag London Limited 2012

In Sect. 6.5, the M-script file that will use the control commands from the joystick as an input to animate the car will be developed.

Section 6.6 will show the reader how to change the **Viewpoint** of the virtual scene. The chapter concludes by a fuzzy logic application for the speed control of the car.

The electronic version of all the M-script files, VRML models, recorded movies for the animated car, in addition to the fuzzy controller for the speed can be downloaded from Springer's web site http://extras.springer.com/.

6.2   Equations of Motion of the Car and Wheels

Given a car that is moving in the xz-plane. The two inputs that are responsible for the planar motion of the car are the traction force F and the steering angle, θ, of the front wheels (the road will be assumed horizontal in the xz-plane). These two inputs are modified by the user instantaneously through the joystick. The inertial coordinate system $\{x, y, z\}$ is centered at O (in VRML, this system has the x- and y-axes to be the horizontal and vertical directions of the computer's screen, respectively). The first body-fixed coordinate system $\{x_1, y_1, z_1\}$ is attached to the car and centered at its center, C (Fig. 6.1), and is fixed to the car. The angle between the x-axis and the x_1-axis is β (Fig. 6.1). Another two body-fixed coordinate systems $\{x_2, y_2, z_2\}$ of center W_{f_l} and $\{x_3, y_3, z_3\}$ of center W_{f_r} are used to describe the orientation and position of the front left and right wheels, respectively. Since both front wheels have the same steering angle θ, the two coordinate systems $\{x_2, y_2, z_2\}$ and $\{x_3, y_3, z_3\}$ have the same unit vectors. The rear left and right wheels are represented by their centers W_{r_l} and W_{r_r} in Fig. 6.1. The degrees of freedom that will be used to model the car and wheels are the horizontal coordinates of point

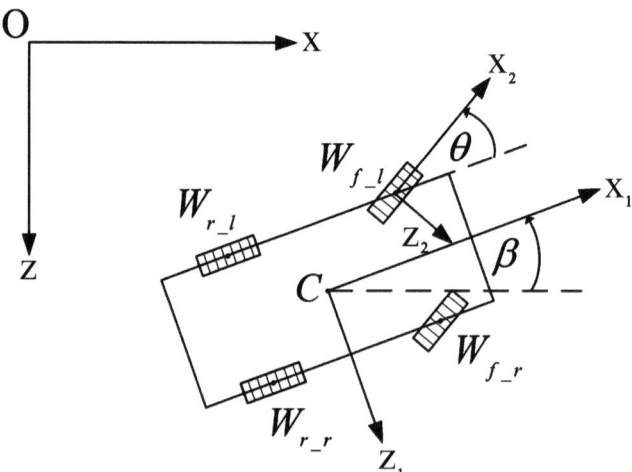

Fig. 6.1 Rotation angles for the car and the steering angle

C (x_C and z_C), orientation of the car, β, in addition to the steering angle of the front wheels, θ.

The velocity of the car in the x_1 direction is V. The latter is denoted by $V(i)$ and $V(i+1)$ at the previous and current time steps, respectively. $V(i+1)$ is proportional to $V(i)$ and to the current traction force $F(i+1)$, and it is computed using the following empirical formula:

$$V(i+1) = 0.3 \times V(i) + F(i+1) \tag{6.1}$$

The velocity of the car, $V(i+1)$, is saturated at its maximum forward value 3.4, and -0.2 m/s in the reverse direction.

The inertial velocity of the car is computed using the rotation matrix between the coordinate systems (x_1, y_1, z_1) and (x, y, z):

$$\begin{bmatrix} V_x(i+1) \\ V_y(i+1) \\ V_z(i+1) \end{bmatrix} = \begin{bmatrix} \cos(\beta(i)) & 0 & \sin(\beta(i)) \\ 0 & 1 & 0 \\ -\sin(\beta(i)) & 0 & \cos(\beta(i)) \end{bmatrix} \begin{bmatrix} V(i) \\ 0 \\ 0 \end{bmatrix} \tag{6.2}$$

Notice that $V_Y(i+1) = 0$ due to the planar motion of the car.

Using the inertial velocity from Eq. 6.2, one can compute the inertial position of the car by computing the discrete integral of the inertial velocity:

$$\begin{bmatrix} x(i+1) \\ y(i+1) \\ z(i+1) \end{bmatrix} = \begin{bmatrix} x(i) \\ y(i) \\ z(i) \end{bmatrix} + \Delta t \times \begin{bmatrix} V_x(i+1) \\ V_y(i+1) \\ V_z(i+1) \end{bmatrix} \tag{6.3}$$

where Δt is the step size.

The orientation of the car, β, is a function of the steering angle θ and the velocity V. $\beta(i+1)$ is computed based on the following empirical formula:

$$\beta(i+1) = \beta(i) + 0.05 \times \theta(i+1) \times V(i+1) + 0.9 \times (\theta(i+1) - \theta(i)) \times V(i+1) \tag{6.4}$$

The current traction force, $F(i+1)$, is computed as follows:

$$F(i+1) = F(i) - 0.05 \times round(joystick\_up\_down) \tag{6.5}$$

where $round(joystick\_up\_down)$ is equal to $+1$ or -1 if the up or down button of the joystick is pressed, respectively.

The current steering angle, $\theta(i+1)$, is computed as follows:

$$\theta(i+1) = \theta(i) - 0.01 \times round(joystick\_left\_right) \tag{6.6}$$

where $round(joystick\_left\_right)$ is equal to $+1$ or -1 if the right or left button of the joystick is pressed, respectively. The lower and upper limits of the steering angle of the car, $\theta(i+1)$, are $-\pi/4$ and $\pi/4$, respectively.

6.3   Creating the Virtual Scene for the Car

The building blocks for our virtual scene are the body of the car (two boxes on top of each other) and the wheels. The body of the car consists of an upper and a lower box that move as one unit (same translation and rotation vectors). The wheels also have the same translation and rotation vectors as the body of the car. In addition, the front wheels steer around their centers (W_{f_l} and W_{f_r} in Fig. 6.1) by the steering angle θ. The dimensions of the car are shown in Fig. 6.2 overlaid on the actual virtual scene. The horizontal plane of the ground is chosen to coincide with the *xz*-plane in VRML. The coordinates of the centers of the upper and lower bodies of the car are (0,1.125,0) and (0,0.375,0), respectively (the *y*-axis in VRML is vertical). The coordinates of the centers of the wheels $W_{r\_l}, W_{r\_r}, W_{f\_l}$ and W_{f_r} are (−0.7,0.2,−0.75), (0.7,0.2,−0.75), (0.7,0.2,0.75), and (−0.7,0.2,0.75), respectively.

Figure 6.3 shows a hierarchy view of the virtual scene of the car. The first level is the virtual scene which is created automatically with each new project. On the second level of the virtual scene, there are four components:

- The **Background**: it is the background of the virtual scene.
- The **transform_car**: it is the **Transform** that will be manipulated in MATLAB® to translate and rotate the car, and thus the **Children** (components) of **transform_car** will be translated and rotated accordingly.
- The **viewpoint_fixed**: it is a fixed **Viewpoint**.
- The **road**: it is a straight road. It is a **Box** component.

On the third level of the virtual scene, there are seven **Children** of **transform_ car**. The **upper_body** and **lower_body** are **Box** components, and they form the body of the car. The **rear_left_wheel** and **rear_right_wheel** are the rear wheels of the car, and they are **Cylinder** components. **Transform_front_left_wheel** and **transform_front_right_wheel** are two **Transform** components that are used as additional layers to rotate the left and right wheels, respectively, using the steering

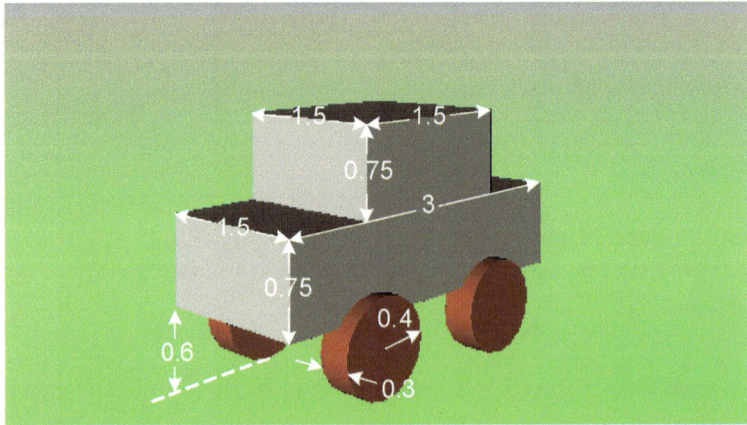

Fig. 6.2 Dimensions of the car

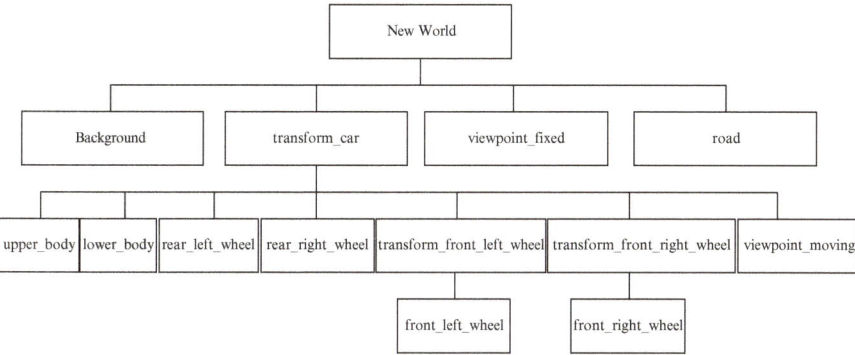

Fig. 6.3 Hierarchy view of the virtual scene of the car

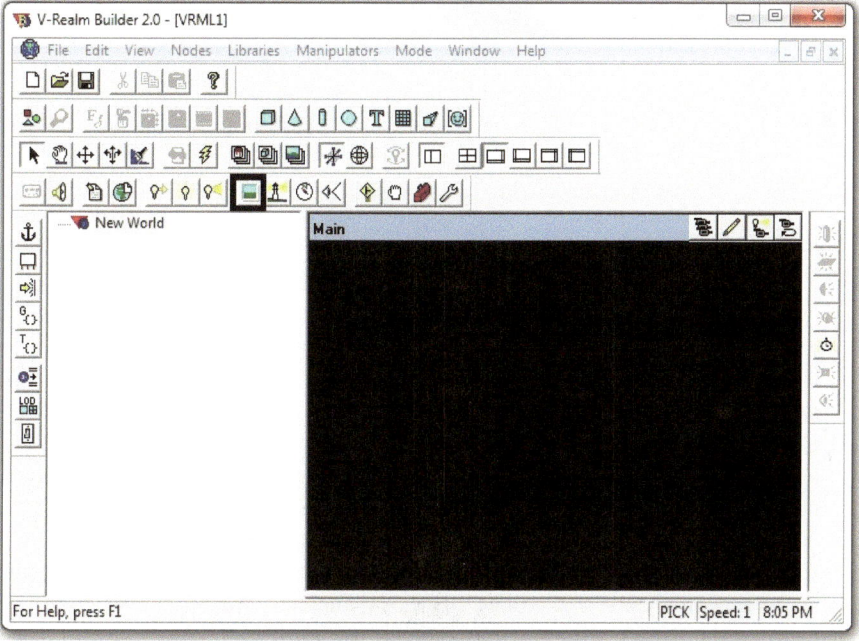

Fig. 6.4 Insert Background button is shown inside the rectangle

angle (these two **Transforms** are optional, and the user could rotate the front wheels without adding them, but for complex systems, transformations become cumbersome; thereby the author advises the user to add the extra **Transforms**). Finally, **viewpoint_moving** is a **Viewpoint** that is moving with the car.

To create the virtual scene for the car, follow the following steps:

1. Open the V-Realm Builder and click on **File>New** to create a new virtual world.
2. Click on the **Insert Background** button (the button is inside the rectangle in Fig. 6.4) to create a background for the scene. The background will be sublayer of **New World**.

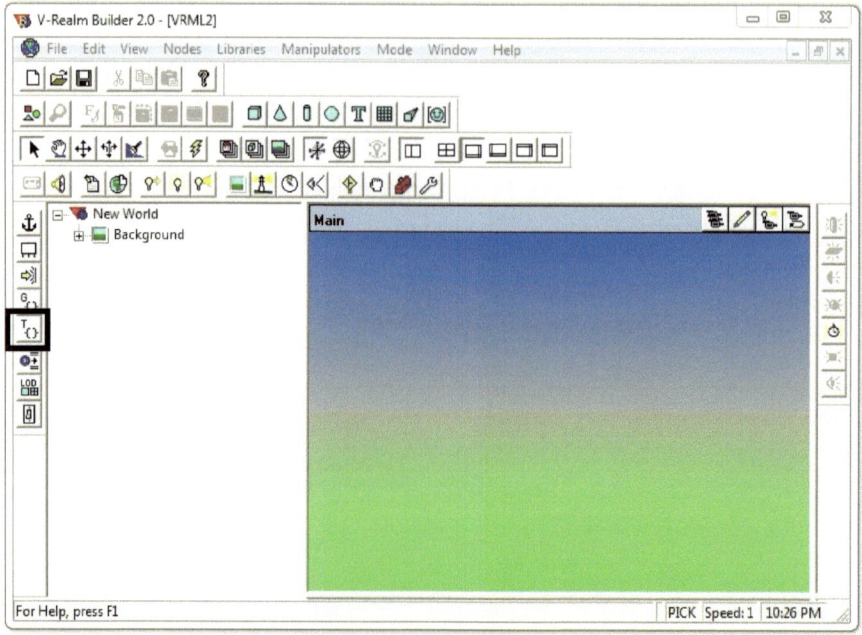

Fig. 6.5   Insert Transform button is shown inside the rectangle

3. Click on **Insert Transform** (Fig. 6.5) to create the **Transform** that will be used
   to translate and rotate the car (body and wheels of the car).
4. Click once on the layer **Transform** to select it. Click another time on **Transform**
   and change the text to a meaningful name such as **transform_car** (Fig. 6.6).
5. Click on the + sign on the left of **transformation_car** to expand all the subtitles
   underneath it. Under the **transform_car** layer, you will have the transformation
   properties such as **translation**, **scaling**, and **rotation** (Fig. 6.7).
6. The last property under **transform_car** is **Children** (Fig. 6.7). Click once on
   Children to select it (Fig. 6.8). This will allow you to add objects (such as the
   body and the wheels) that will be a part of **transform_car**.
7. Click on the **Insert Box** (Fig. 6.9) to create the upper body of the car.
8. Click once on the **Transform** layer to select it. Click another time on **Transform**
   and change the name to a meaningful name such as **upper_body** (Fig. 6.10).
9. To resize the box of the **upper_body** to the dimensions provided in Fig. 6.2,
   click on the + sign on the left of **Box** to expand all the layers underneath it.
   Under the **Box** icon, you will have the size property of the box of **upper_body**
   (Fig. 6.11).
10. Double-click the size property of the **upper_body** and change the dimensions
    by ticking the boxes beside the x-axis, y-axis, and z-axis and changing the val-
    ues to 1.5, 0.75, and 1.5, respectively. Press OK when done (Fig. 6.12).
11. Double-click on the **translation** property to translate **upper_body** (Fig. 6.13).
    The box should be translated by 1.725 in the y-direction (the vertical direction).

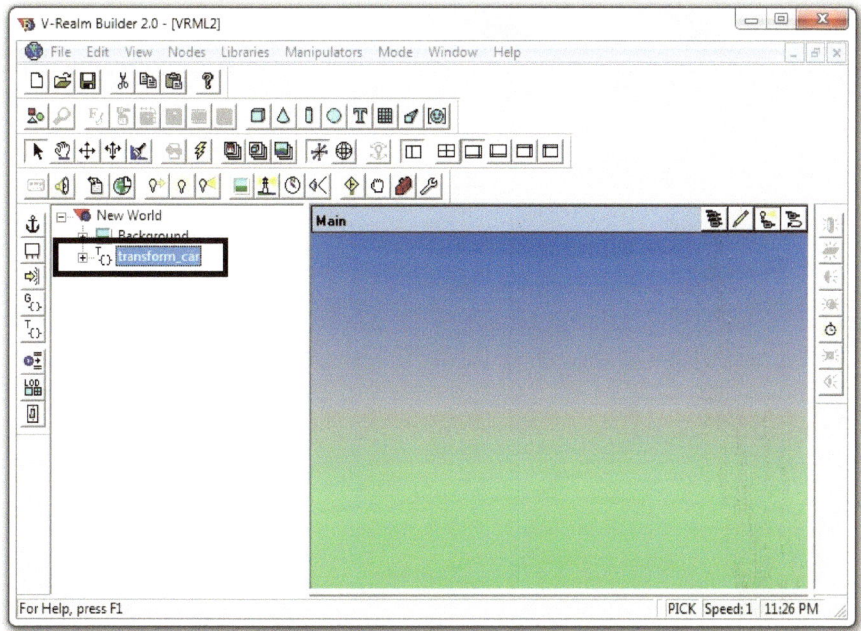

Fig. 6.6 Change the name of the Transform to transform_car

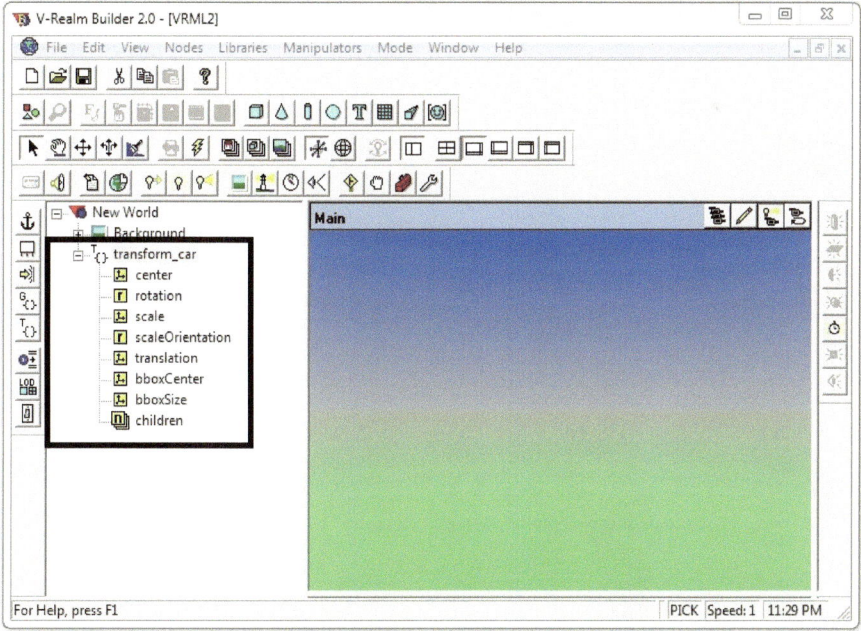

Fig. 6.7 Transformation properties

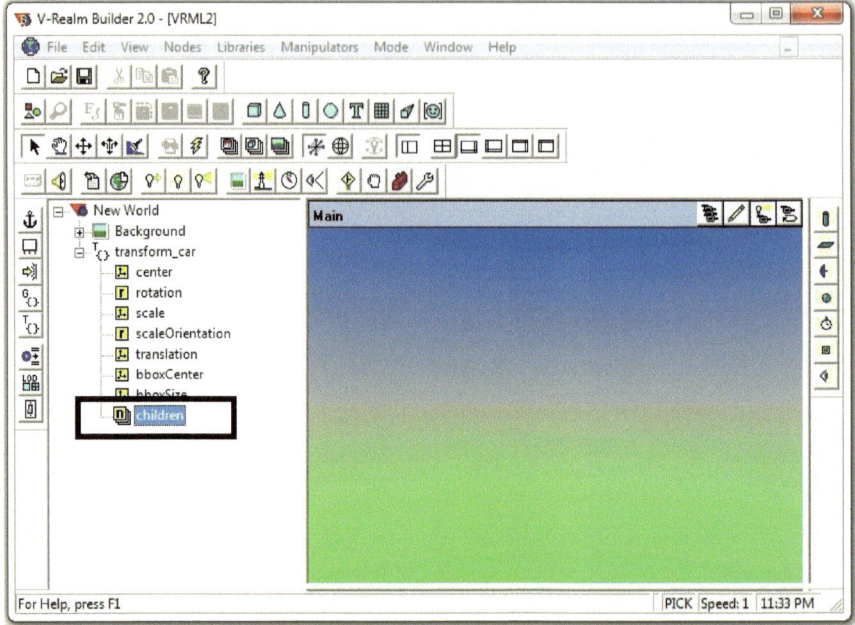

Fig. 6.8   The Children property of transform_car

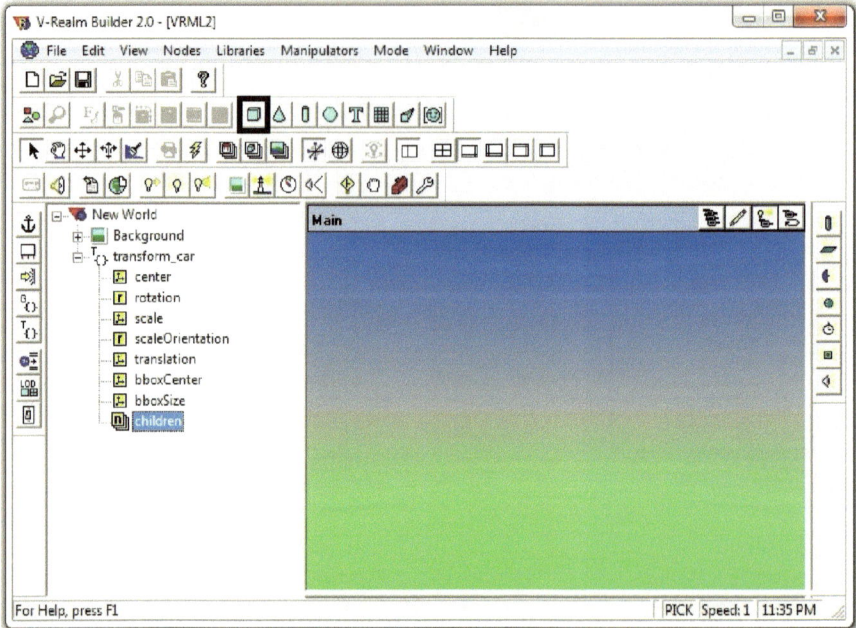

Fig. 6.9   Insert Box button is shown inside the rectangle

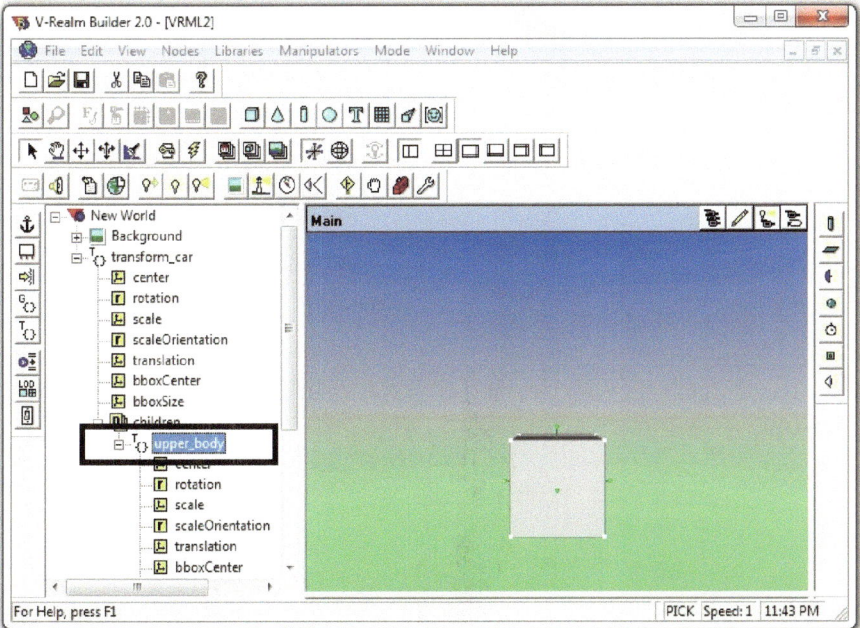

Fig. 6.10 Rename Transform to upper_body

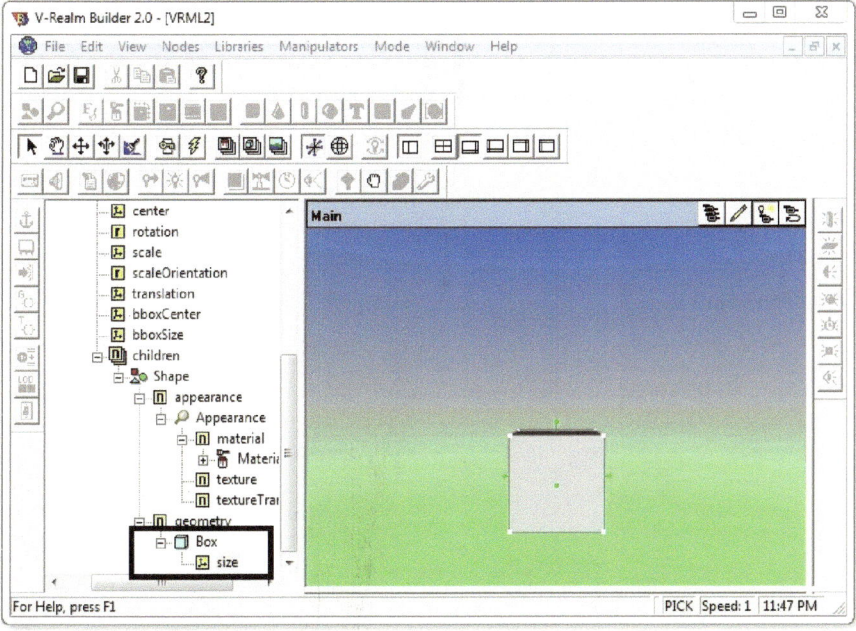

Fig. 6.11 Size property of Box

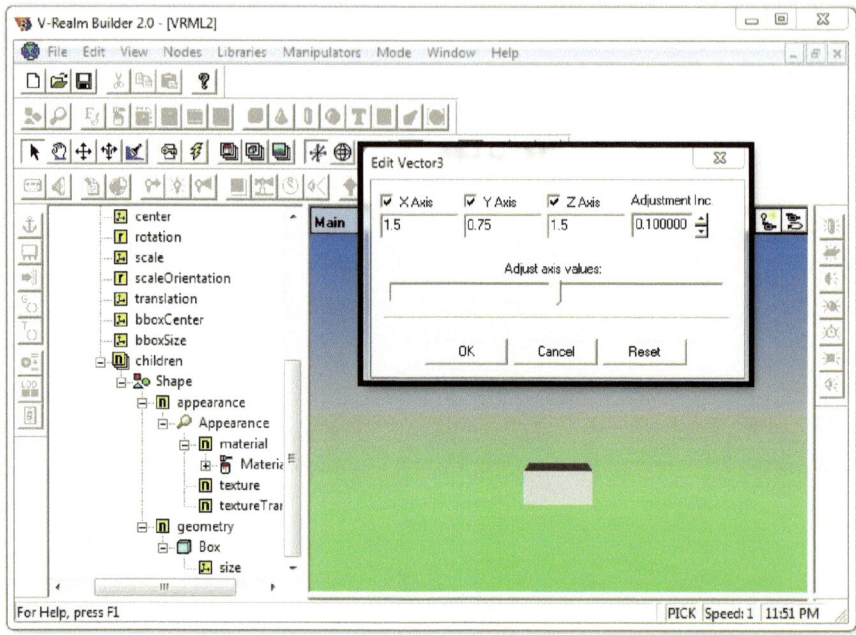

Fig. 6.12   Resize the Box

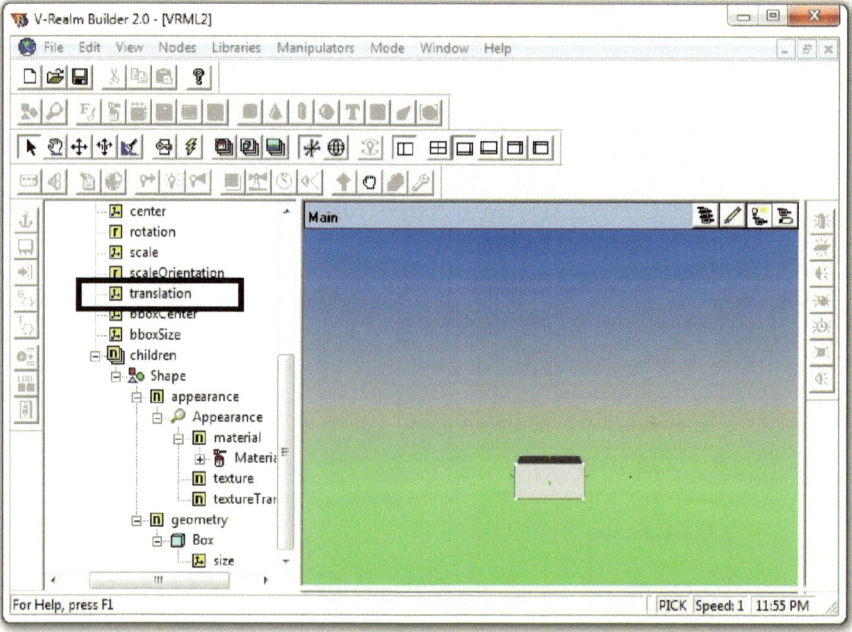

Fig. 6.13   The translation property of the upper_body

This value (1.725) is equal to the distance from ground (0.6) plus the height of the lower body (0.75) plus half the height of the upper body (0.75/2 is where the center of the box is located). Refer to Fig. 6.2 for these dimensions. Press OK when done translating the **upper_body**.

12. Repeat steps 6–8 to create the lower body. Name the **Transform lower_body**.
13. Repeat steps 9 and 10 to resize the dimensions of the lower body to 1.5, 0.75, and 3 in the *x*-, *y*-, and *z*-directions, respectively. Refer to Fig. 6.2 for these dimensions.
14. Repeat step 11 to translate the lower body by 0.955 in the *y*-direction.
15. Repeat step 6 to select **Children** layer of **transform_car**.
16. Click on the **Insert Cylinder** (Fig. 6.14) to create the rear left wheel.
17. Click once on the new **Transform** to select it. Click another time on **Transform** and change the text to a meaningful name such as **rear_left_wheel**. Press Enter when done.
18. To change the dimensions of the **rear_left_wheel**, click on the + sign on the left of **Cylinder**, thus expanding all the layers underneath it. You will see the properties of the **Cylinder** such as the radius and height (Fig. 6.15).
19. Change the height of the **Cylinder** of **rear_left_wheel** to 0.3 by double-clicking on the height property and changing its value in the box. Press OK when done. Similarly, change the radius to 0.4. These dimensions are provided in Fig. 6.2.
20. From Fig. 6.16, you can see that the **rear_left_wheel** needs to be rotated about its *z*-axis by 90°.

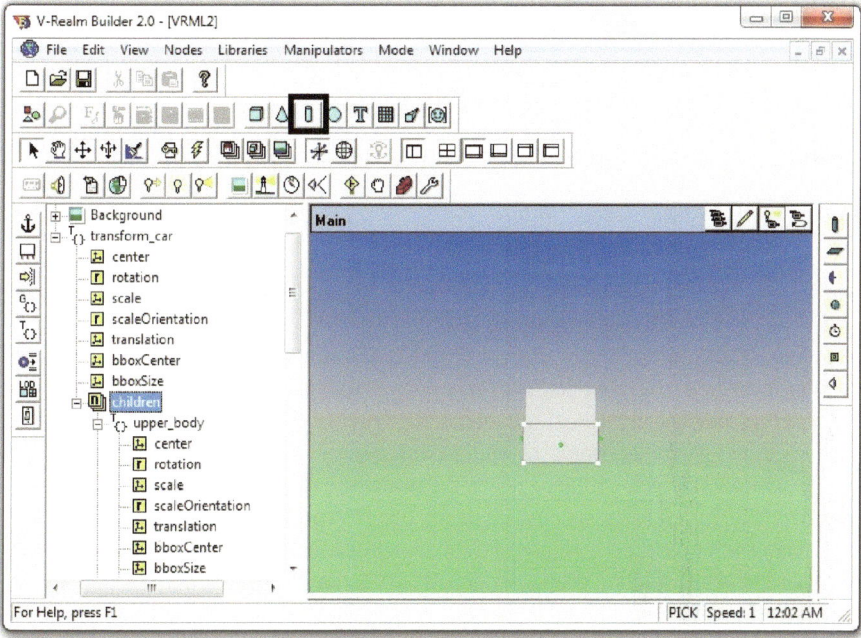

Fig. 6.14 Insert Cylinder button is shown inside the rectangle

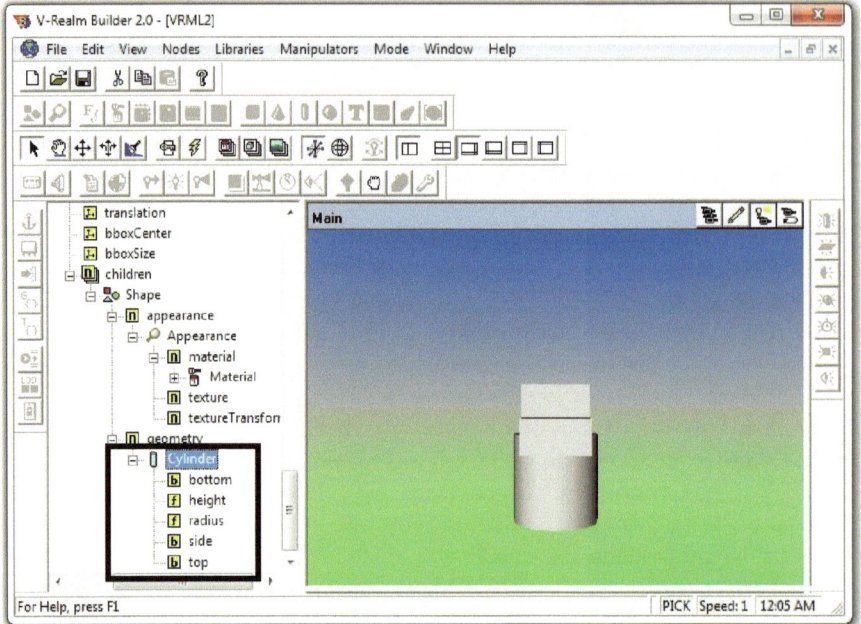

Fig. 6.15 Properties of Cylinder

21. Rotate the **rear_left_wheel** by double-clicking the **rotation** property (Fig. 6.16). The values under the *x*-, *y*-, and *x*-axes should be 0, 0, and 1 so that the **rear_left_wheel** would rotate about a unit vector in the *z*-direction. Change the **rotation** value to 90° (Fig. 6.17). Press OK when done.

22. Double-click on the **translation** property to translate the **rear_left_wheel** (Fig. 6.18). The wheel should be translated by −0.7, 0.4, and −0.75 in the *x*-, *y*-, and *z*-directions, respectively.

23. Repeat steps 15–21 to create the rear right wheel. Change the name of the created Transform to **rear_right_wheel**.

24. Repeat step 22 to translate the **rear_right_wheel** by 0.7, 0.4, and −0.75 in the *x*-, *y*-, and *z*-directions, respectively.

25. The creation of the front wheels differs from those in the back by the fact that the front ones will rotate about their vertical axis by the steering angle θ. Thus, to create each front wheel, an additional layer of transform will be added, and this transformation should be under **Children** of **transform_car**. Repeat step 6 to select **Children** sublayer under **transform_car**.

26. To create the left front wheel, click on the **Insert Transform** (Fig. 6.5). This transformation will hold the translation and rotation properties of the right front wheel (these translation and rotation properties do not vary as the steering θ varies). Rename the **Transform** to a meaningful name such as **transform_front_left_wheel**.

27. Click on the + sign on the left of **transform_front_left_wheel** to expand all the properties underneath it. Repeat step 21 to change the rotation property of **transform_front_left_wheel**.

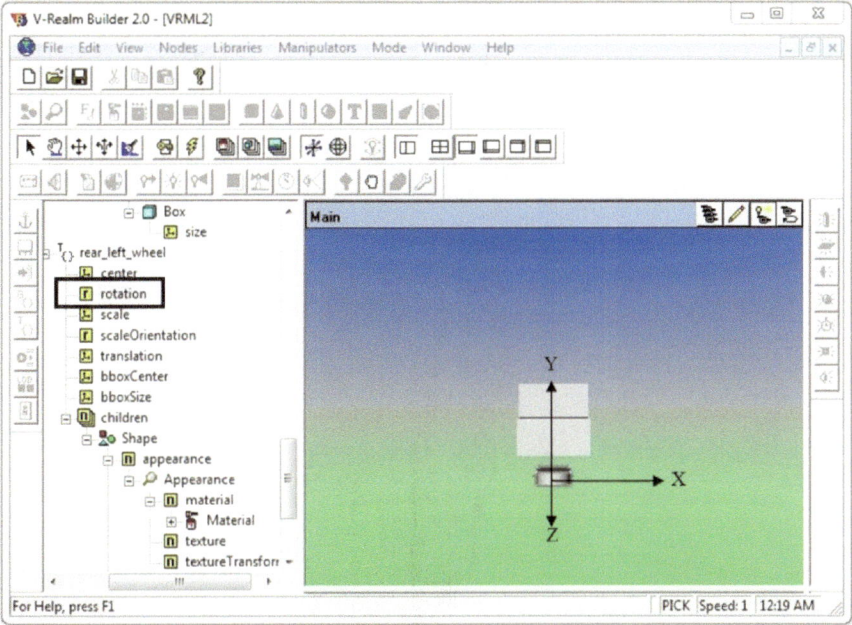

Fig. 6.16   The rotation property of the rear_left_wheel in the rectangle (on left) in addition to the axes of rotation of the cylinder (on right)

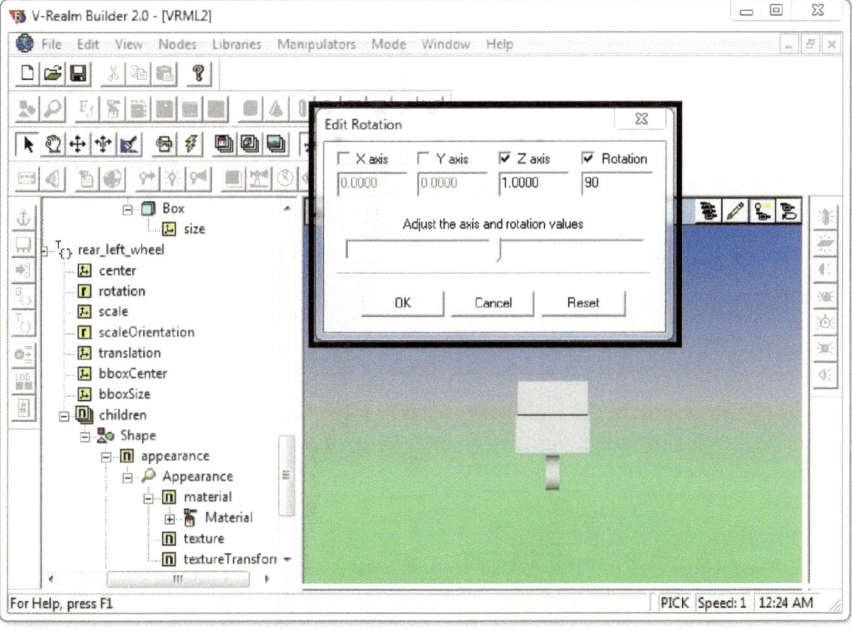

Fig. 6.17   Rotation of the rear_left_wheel

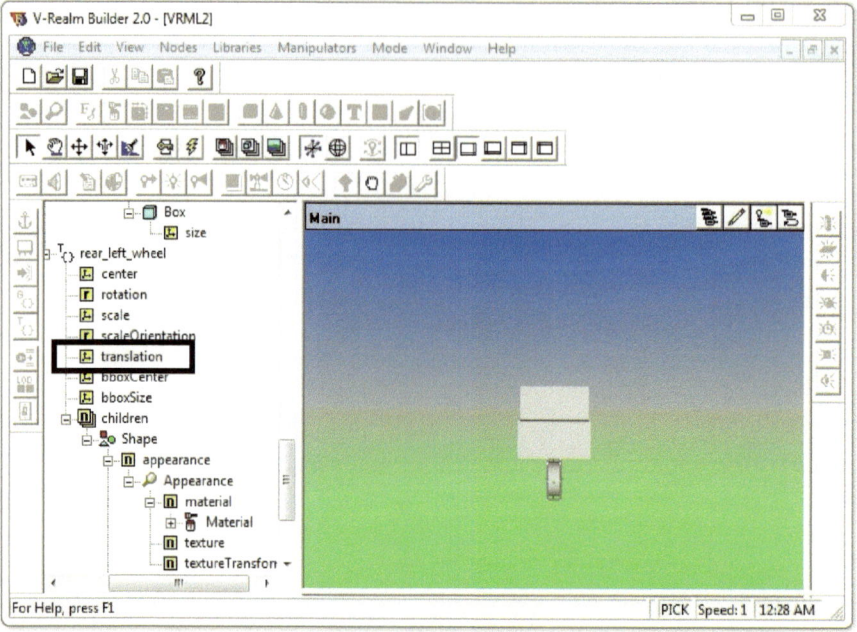

Fig. 6.18 The translation property of rear_left_wheel is shown in the rectangle

28. Repeat step 22 to change the **translation** property of **transform_front_left_wheel** to 0.7, 0.4, and 0.75 in the *X*-, *Y*-, and *Z*-directions, respectively.

29. The last property of **transform_front_left_wheel** is **Children**. Click once on **Children** to select it. This will allow you to add the front left wheel under the modified translation and rotation properties. Click on the **Insert Cylinder** (Fig. 6.14) to create the front left wheel.

30. Click once on the new title **Transform** to select it. Click another time on **Transform** and change the text to a meaningful name such as **front_left_wheel**. Press Enter when done.

31. To change the dimensions of the **front_left_wheel,** click on the + sign on the left of **Cylinder**, thus expanding all the sublayers underneath it. These are the properties of the **Cylinder,** and they include the radius and height (Fig. 6.15).

32. Change the height of the **Cylinder** of **front_left_wheel** to 0.3 by double-clicking on the **height** property. Similarly, change the radius to 0.4 by double-clicking the **radius** properties, respectively.

33. To create the front right wheel, repeat steps 25 and 26. Rename the transform **transform_front_right_wheel**.

34. Click on the + sign on the left of **transform_front_right_wheel** to expand all the properties underneath it. Repeat step 21 to change the rotation property of **transform_front_right_wheel.**

35. Repeat step 22 to change the translation property of **transform_front_right_wheel** to −0.7, 0.4, and 0.75 in the *X*-, *Y*-, and *Z*-directions, respectively.

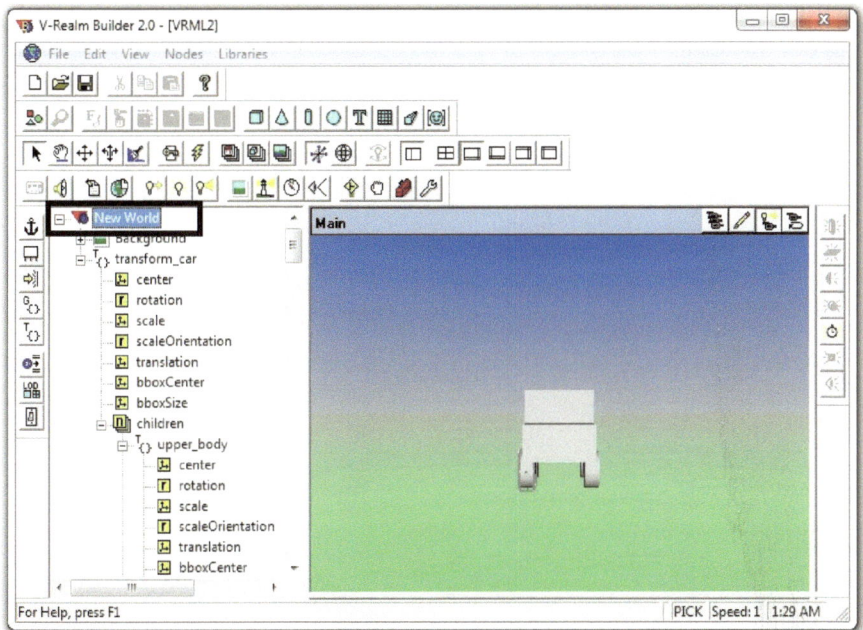

Fig. 6.19 Top level of the virtual scene

36. The last property of **transform_front_right_wheel** is **Children**. Click once on **Children** to select it. This will allow you to add the front right wheel under the modified translation and rotation properties. Click on the **Insert Cylinder** (Fig. 6.14) to create the front right wheel.

37. Click once on the new title **Transform** to select it. Click another time on **Transform** and change the text to a meaningful name such as **front_right_wheel**. Press Enter when done.

38. To change the dimensions of the **front_right_wheel**, click on the + sign on the right of **Cylinder**, thus expanding all the sublayers underneath it. These are the properties of the **Cylinder**, and they include the radius and height (Fig. 6.15).

39. Change the height of the **Cylinder** of **front_right_wheel** to 0.3 by double-clicking on the **height** property. Similarly, change the radius to 0.4 by double-clicking the **radius** properties, respectively.

40. In the following steps, the straight road will be created. To do so, choose the top level of the virtual scene (Fig. 6.19); this will allow you to add an object that is not a child of any of the previous objects or transforms in the virtual scene.

41. Click on the **Insert Box** (Fig. 6.9) to create the road. Rename the created **Transform** to a meaningful name such as **road**. Press Enter when done.

42. Repeat steps 9 and 10 to resize the Box property of **road** to 4, 0.1, and 200 in the *X*-, *Y*-, and *Z*-directions, respectively.

43. Select the top level of the virtual scene (Fig. 6.20).

44. The point where the viewer will be looking at the scene from is called a **Viewpoint**. To create a **Viewpoint**, click Nodes>Insert>Bindable>Viewpoint (Fig. 6.21).

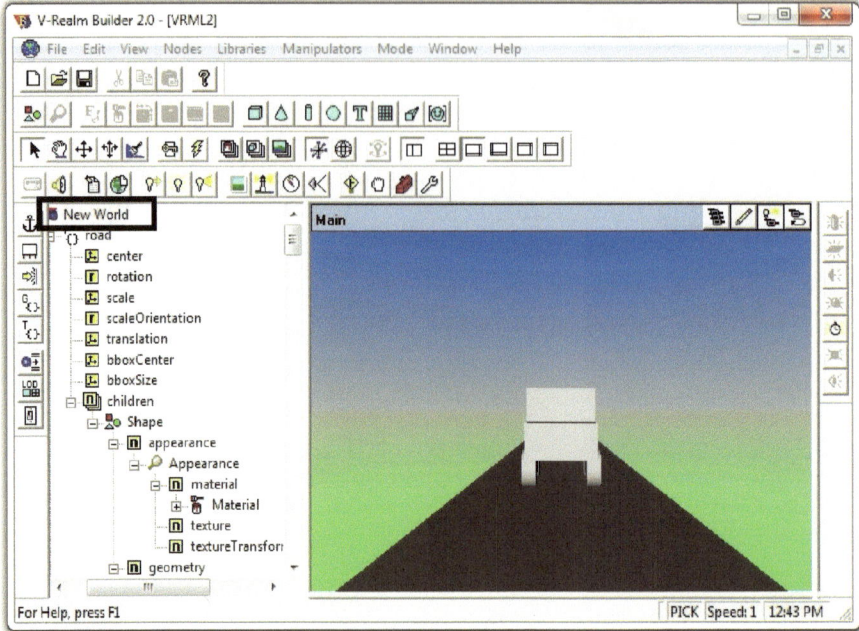

Fig. 6.20   Top level of the virtual scene

45. Rename the newly created **Viewpoint viewpoint_fixed**.
46. Click on the + sign on the left of **viewpoint_fixed** to expand all the subtitles underneath it. Double-click the **description** property of **viewpoint_fixed** (Fig. 6.22) and type **viewpoint_fixed**. The text in the description property is the name that MATLAB® will uniquely identify in this viewpoint.
47. Double-click the **position** property of **viewpoint_fixed** (Fig. 6.23) and change the *Y*-axis value to 2. This will increase the elevation from which the viewer, when situated at **viewpoint_fixed**, sees the view.
48. To actually observe the view from the current viewpoint after the changes and check if the position of the current **Viewpoint** is acceptable, double-click the **set_bind** property (Fig. 6.24) and change its value to True (this step is not mandatory, and it is only done for checking the **Viewpoint**).
49. To create a **Viewpoint** that will be moving with the car, select the **Children** layer of **transform_car** and repeat step 44.
50. Rename this viewpoint **viewpoint_moving**.
51. Repeat step 46 to type the **description** property of the new viewpoint. Fill the description as **viewpoint_moving**.
52. Repeat step 47 to change the **position** property (Fig. 6.23) of **viewpoint_moving**. Change the *Y*-axis value to 3.
53. Repeat step 48 to check the new position of the Viewpoint.
54. Save the final virtual scene by clicking on **File>Save As**. Name the virtual word *car*.

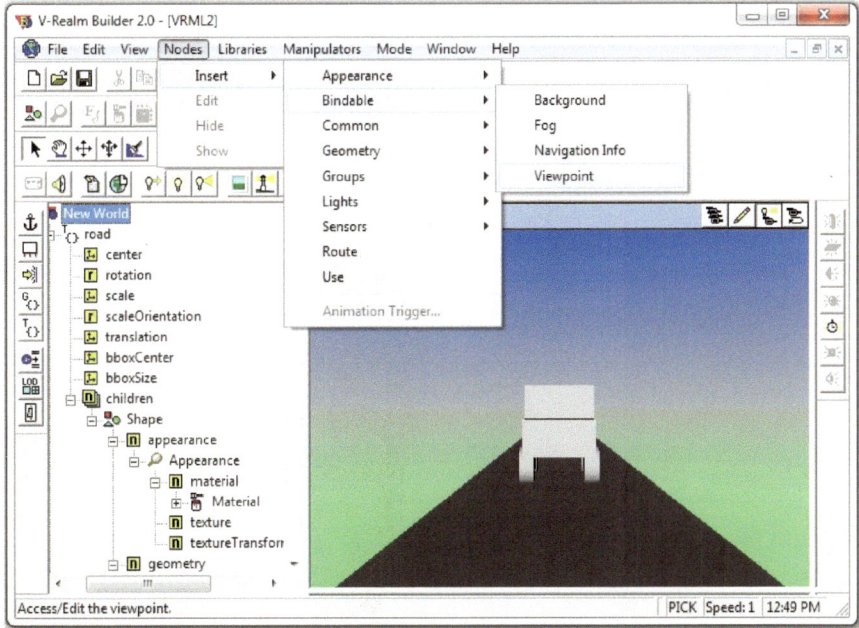

Fig. 6.21   Nodes>Insert>Bindable>Viewpoint

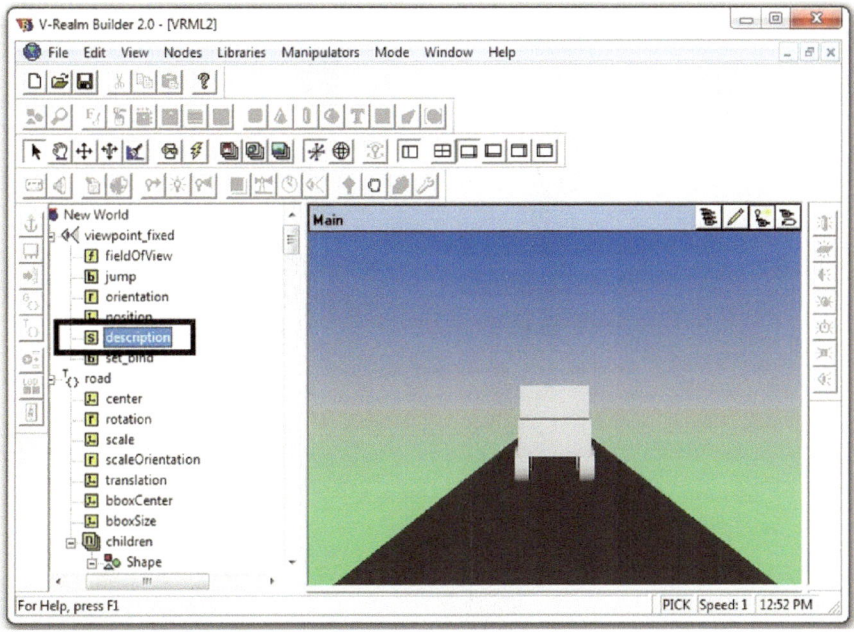

Fig. 6.22   The description property of viewpoint_fixed

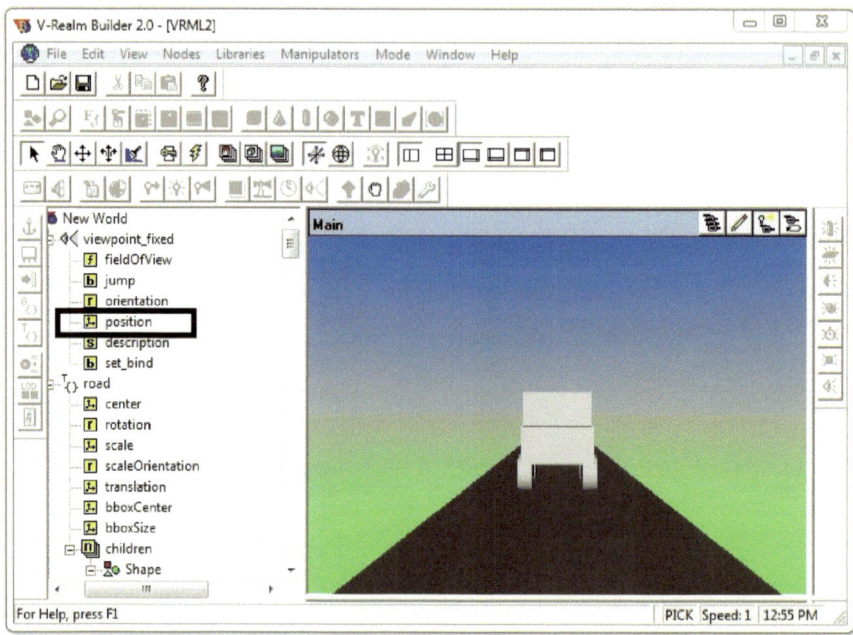

Fig. 6.23 The position property of viewpoint_fixed

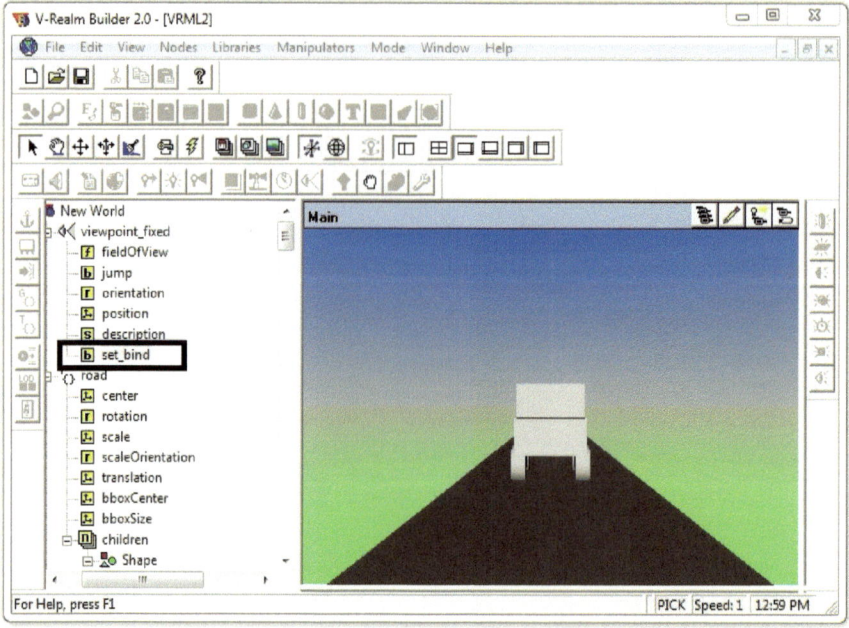

Fig. 6.24 The set_bind property of viewpoint_fixed

If you want to add a texture to or color some of the parts of the car, please refer to Sect. 4.4.

6.4   Joystick Setup

Any simple USB joystick could be used along with this example. The one that has been used to run this example is "Logitech Precision Joystick" (Fig. 6.25). It does not require any setup. It should be a plug-and-play device, and it is recommended to be plugged in the USB port before starting the MATLAB® session. The buttons that will be used in running this example in MATLAB® are shown in Fig. 6.25.

6.5   M-script File for Animating the Virtual Scene

In what follows, the virtual scene, *car.wrl*, will be animated by varying the traction force and the steering angle by using the joystick. The variables that will be changed in the virtual world are:

1. The **translation** property of **transform_car** (which represents the translation of the car's body with the wheels)
2. The **rotation** property of **front_right_wheel** (which represents the rotation of the front right wheel by the steering angle, θ)
3. The **rotation** property of **front_left_wheel** (which represents the rotation of the front left wheel by the steering angle, θ)

Fig. 6.25 "Logitech Precision Joystick" with the required buttons for this chapter

    The following is the M-script that is used to animate the virtual model of the car. The comments should give the reader a good understanding of how to animate the virtual scene with the inputs from the joystick.

car_animate.m

```
%%%%%%%%%%%%%%%%%%%%%%%%%%%%%%%%%%%%%%%%%%%%
% Initialize the virtual model
%%%%%%%%%%%%%%%%%%%%%%%%%%%%%%%%%%%%%%%%%%%%
% Create a virtual world associated with car.wrl file
world = vrworld('car.WRL');
% Open the virtual world
open(world);
% Draw the virtual world
fig = view(world, '-internal');


%%%%%%%%%%%%%%%%%%%%%%%%%%%%%%%%%%%%%%%%%%%%
%%%%  Parameters of the model %%%%%%
%%%%%%%%%%%%%%%%%%%%%%%%%%%%%%%%%%%%%%%%%%%%
% Mass of the car
m=1000;
% Time step
delta_t=0.1;


%%%%%%%%%%%%%%%%%%%%%%%%%%%%%%%%%%%%%%%%%%%%
%%%%%%%  Initial conditions %%%%%%%%
%%%%%%%%%%%%%%%%%%%%%%%%%%%%%%%%%%%%%%%%%%%%
% Initial position of the car
Position_Car_i=[0;0;0];
% Initial velocity of the car
V_i=0;
% Initial orientation of the car
betta_i=pi;
% Initial traction force
F_i=0;
% Initial steering angle
theta_i=0;


%%%%%%%%%%%%%%%%%%%%%%%%%%%%%%%%%%%%%%%%%%%%
%%%%%%%  Define the joystick %%%%%
%%%%%%%%%%%%%%%%%%%%%%%%%%%%%%%%%%%%%%%%%%%%
%% 1 is the default number  for the joystick.
%You have to change it in case you have more than 1
%joystick
joy = vrjoystick(1);
```

```
%%%%%%%%%%%%%%%%%%%%%%%%%%%%%%%%%%%%%%%%%%
%%% for loop to run the script  %%%
%  across the entire desired time%%
%%%%%%%%%%%%%%%%%%%%%%%%%%%%%%%%%%%%%%%%%%
for t=0:0.1:100
   %Use the pause command to artificially slow
   % the simulation run time close to real time
   pause(.05);

   % Define the joystick axis:
   %    a(1) represents the values from
   %    the left and right direction buttons
   %    where
   %    a(1)~ -1 for left direction
   %    a(1)~ 1 for right direction
   %    a(2) represents the values from
   %    the up and and down direction buttons
   %    where
   %    a(2)~-1 for up
   %    a(2)~1 for down
   a = axis(joy, [1 2]);

   %Compute the steering angle theta using eq (6.6) by
   % adding the previous angle to an increment of
   % +/- 0.01 based on the pressed left/right
   % joystick button. round function is used since
   %the values returned from the joystick are close to
   %1 in magnitude but might not be exactly 1
   theta_i1=theta_i-0.01*round(a(1));

   %%Limit the value of the steering angle between -pi/4 and pi/4
   if theta_i1>pi/4
      theta_i1=pi/4;
   elseif theta_i1<-pi/4
      theta_i1=-pi/4;
   end

   %Compute the traction force based on eq (6.5)
   % add the previous angle to an increment of
   % +/- 0.05 based on the pressed up/right
   % joystick button . round function is used since
   %the values returned from the joystick are close to
```

```
%1 in magnitude but might not be exactly 1
F_i1=F_i-0.05*round(a(2));

%Solve for eq (6.1) to compute the forward speed
%of the car
V_i1=0.3*V_i+F_i1;

%Limit the values of forward and backward speed
if V_i1>3.4
    V_i1=3.4;
elseif V_i1<-0.2
    V_i1=-0.2;
end

%Solve eq (6.2) to compute the inertial velocity of the car
Velocity_Car1=[cos(betta_i) 0 sin(betta_i);0 1 0;-sin(betta_i) 0
cos(betta_i)]*[0;0;V_i1];

%Solve eq (6.3) to comppute the inertial position of the car
Position_Car_i1=Position_Car_i+delta_t.*Velocity_Car1;

%Solve eq (6.4)  to compute the orientation of the car
betta_i1=betta_i+0.05*theta_i1*V_i1+0.9*(theta_i1-theta_i)*V_i1;

%Change the translation property of transform_car to translate the car
world.transform_car.translation=(Position_Car_i1)';

%Change the rotation property of transform_car to rotate the care
world.transform_car.rotation=[0 1 0 betta_i1];

%Change the rotation property of front_right_wheel
world.front_right_wheel.rotation=[1 0 0 theta_i1];

%Change the rotation property of front_left_wheel
world.front_left_wheel.rotation=[1 0 0 theta_i1];

%Update the states according to the current computed values
Position_Car_i=Position_Car_i1;
V_i=V_i1;
betta_i=betta_i1;
theta_i=theta_i1;
F_i=F_i1;
%Update the virtual scene
vrdrawnow;
end
```

6.6   Changing the Viewpoint

Two **Viewpoints** have been created for the animation: **viewpoint_fixed** and **viewpoint_moving**. To change the **Viewpoint**, the top section of the code *car_animate.m* that initializes the virtual model has to be evaluated:

```
%%%%%%%%%%%%%%%%%%%%%%%%%%%%%%%%%%%%%%%%%%%%%%%%%%%%%%
%  Initialize the virtual model
%%%%%%%%%%%%%%%%%%%%%%%%%%%%%%%%%%%%%%%%%%%%%%%%%%%%%%
% Create a virtual world associated with car.wrl file
world = vrworld('car.WRL');
% Open the virtual world
open(world);
% Draw the virtual world
fig = view(world, '-internal');
```

Once this part is evaluated, the virtual scene of the car will be opened. The drop menu in Fig. 6.26 can be used to change the **Viewpoint**. Once the desired **Viewpoint** is chosen, the rest of the script can be evaluated to animate the car with the new **Viewpoint**.

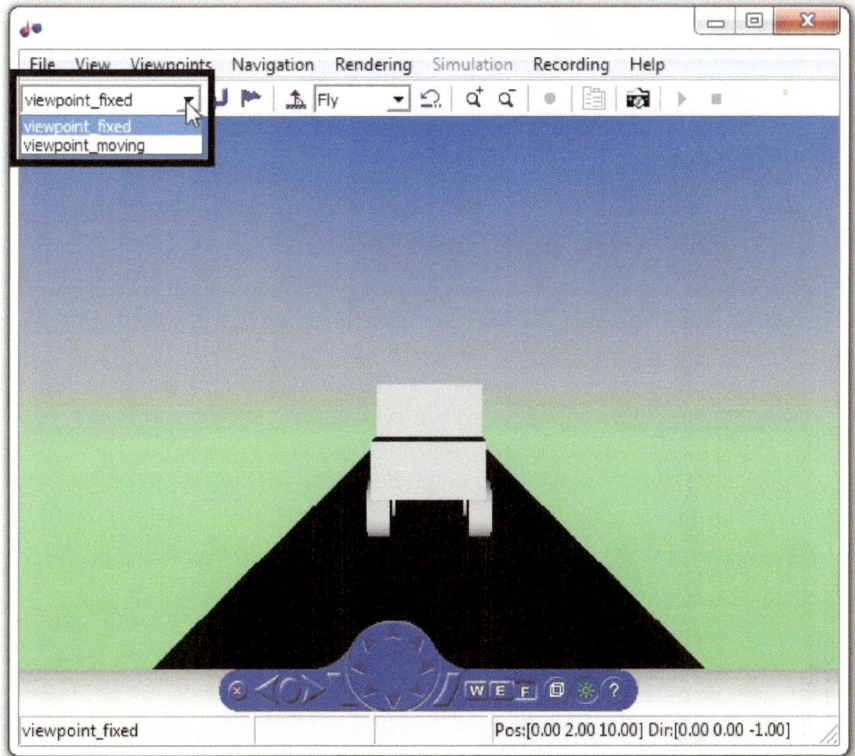

Fig. 6.26 Drop-down menu to change the Viewpoint

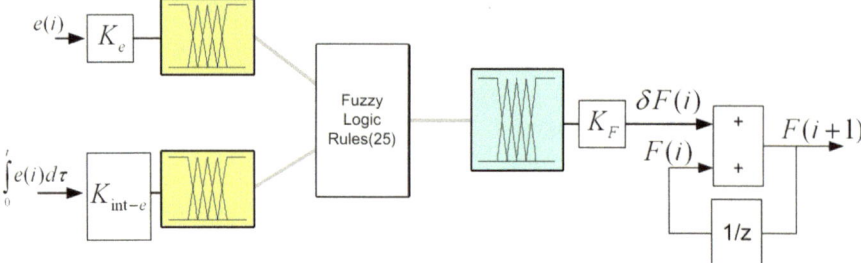

Fig. 6.27 Incremental fuzzy logic speed controller

Table 6.1 Rules for the incremental fuzzy logic controller

$\int e$	e				
	NL	NS	Z	PS	PL
NL	NL	NL	NS	PS	PL
NS	NL	NS	Z	PS	PL
Z	NL	NS	Z	PS	PL
PS	NL	NS	Z	PS	PL
PL	NL	NS	PS	PL	PL

NL stands for negative large, *NS* stands for negative small, *Z* stands for zero, *PS* stands for positive small and *PL* stands for positive large

6.7 Application Problem: Fuzzy Logic Controller for the Car Speed

The main objective of this section is to design a speed controller for the car model presented in Sect. 6.2. In this problem, the car steering angle will still be controlled by the joystick, but the traction force, $F(i+1)$, will be commanded by the controller to maintain a fixed speed of 3 m/s for the car. The car should reach the desired speed in a maximum time of 3 s with zero overshoot. The fuzzy logic controller (Yen and Langari 1999) is incremental (i.e., the value of $F(i+1) = F(i) + \delta F(i)$, where δF is the current output of the fuzzy logic controller). The structure of the fuzzy logic controller is explained in Fig. 6.27. e is the speed error and is defined as $V_{\text{target}}(i) - V(i) = 3 - V(i)$. K_e, $K_{\text{int}-e}$, and K_F are the gains of the controller. The 25 rules for the fuzzy logic controller are summarized in Table 6.1.

Figures 6.28, 6.29, and 6.30 show the plots for the membership functions for $e(i)$, $\int_0^t e(i)d\tau$, and $\delta F(i)$. Solve the following:

1. Using the MATLAB® Fuzzy Logic Toolbox (type in **fuzzy** in the MATLAB® workspace), construct the speed fuzzy controller based on the rules given in Table 6.1 and the membership functions in Figs. 6.28, 6.29, and 6.30.

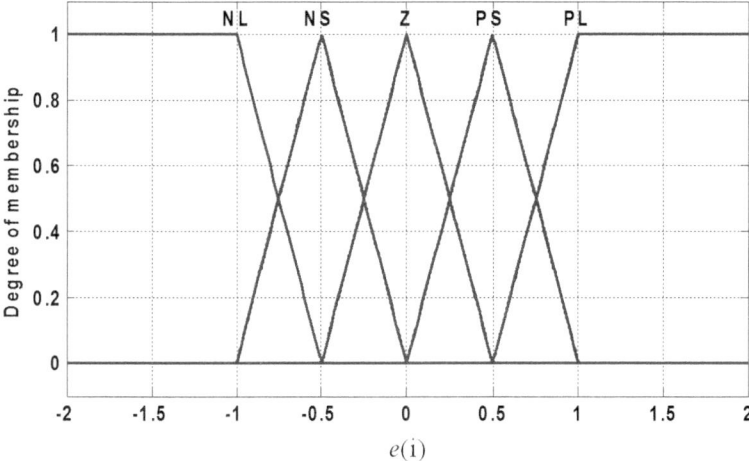

Fig. 6.28 Membership functions for the error $e(i)$

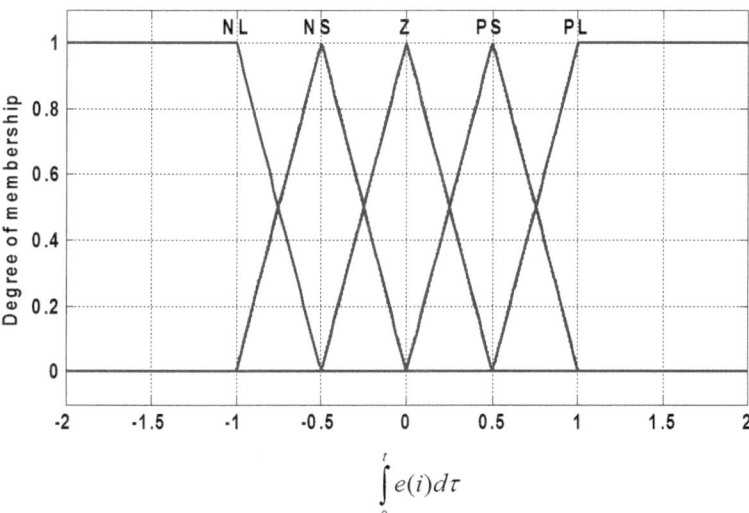

Fig. 6.29 Membership functions for the integral of the error $\int_0^t e(i)d\tau$

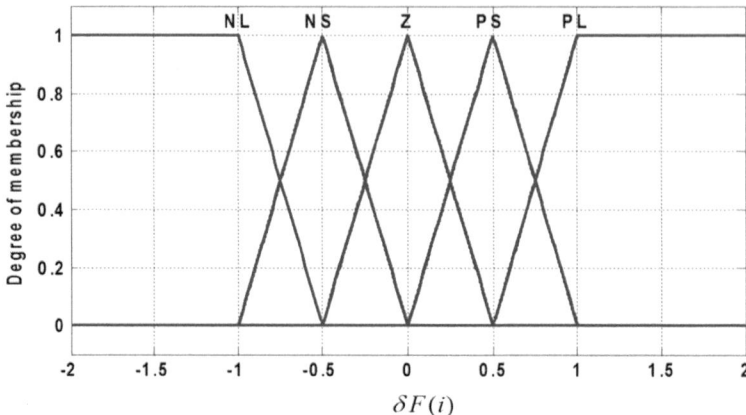

Fig. 6.30 Membership functions for the increment of the controller output

2. Use the developed controller along with the M-script file generated to animate the car (***car_animate.m***) to animate the controlled speed of the car in VRML. Plot the instantaneous and desired speeds of the car while running the simulation.
3. Tune the controller by playing around with the gains K_e, K_{int-e}, and K_F in order to meet the requirements.

The solution of the problem can be downloaded from Springer's web site http://extras.springer.com/. The files can be found in folder /Chapter 6/Application Problem.

Reference

Book

Yen J, Langari R (1999) Fuzzy logic: intelligence, control and information. Prentice Hall, Upper Saddle River

Chapter 7
Animation of a Ship Moving Across Waves

7.1 Introduction

This chapter aims to teach the reader how to handle meshing and animation of a 3D surface like the sea surface using MATLAB®. In addition, the reader will learn how to create and handle **Spot Lights** in VRML. Furthermore, the user will learn how to add readily available objects from the library of VRML to the virtual scene. The reader will also implement rotations and translations in VRML to animate a ship moving across sea waves.

In Sect. 7.2, the governing equations of motion of the waves and the ship will be introduced.

In Sect. 7.3, a step-by-step guide to draw the virtual scene for the waves and the ship in VRML will be presented. In addition, a **Spot Light** will be added to the scene to help illuminate all the components.

In Sect. 7.4, the M-script that will be used to animate the scene will be introduced.

The chapter concludes by an application problem that includes drawing a virtual scene in VRML similar to the one developed in Sect. 7.3 in addition to developing a fuzzy logic controller for the heading of the ship.

The electronic version of all the M-script files, VRML models, fuzzy controller for the heading, and video recordings for the animations can be downloaded from Springer's web site http://extras.springer.com/.

7.2 Equations of Motion of Sea Waves and the Ship

In general, sea waves are modeled as short or long crested (Perez 2005; Khaled 2010). In this chapter, waves will be modeled as sinusoidal with one frequency ω. These waves will be traveling in the z-direction, and the wave height at a certain point will be computed based on Eq. 7.1:

Matlab® is a registered trademark of The Mathworks, Inc.

N. Khaled, *Virtual Reality and Animation for MATLAB® and Simulink® Users:*
Visualization of Dynamic Models and Control Simulations,
DOI 10.1007/978-1-4471-2330-9_7, © Springer-Verlag London Limited 2012

$$h(t, z) = 2\sin(\omega t + z) \tag{7.1}$$

The ship will be moving along the x-direction, starting from the initial position $x(0) = -100$, according to Eq. 7.2:

$$x(t) = -100 + 10t \tag{7.2}$$

7.3   Creating the Virtual Scene for the Sea and Ship

The building blocks for the virtual scene are the sea waves, the ship, and the background of the mountains (Fig. 7.1). In addition, the scene will include a **Viewpoint** and a **Spot Light** (a source of light shed on the scene).

Some of the figures that will be shown in this chapter might not render in VRML properly based on the specifications of the used computer, but they should render when the virtual scene is called by MATLAB® and Simulink®. The grid of the sea surface is one of the components that might not render properly especially if the number of grids is too high. Thus, if it does not show in the virtual scene, or if it appears as a white plane, this should not cause a problem in MATLAB® and Simulink®. To create the virtual scene for the sea, follow the following steps:

1. Open the V-Realm Builder and click on **File>New** to create a new virtual world.
2. Click on the **Insert Background** button (Fig. 7.2) to create a background for the scene. The background will be sublayer of **New World**.
3. Click on the **Insert Elevation Grid** button (Fig. 7.3) to create the grid of the sea surface.

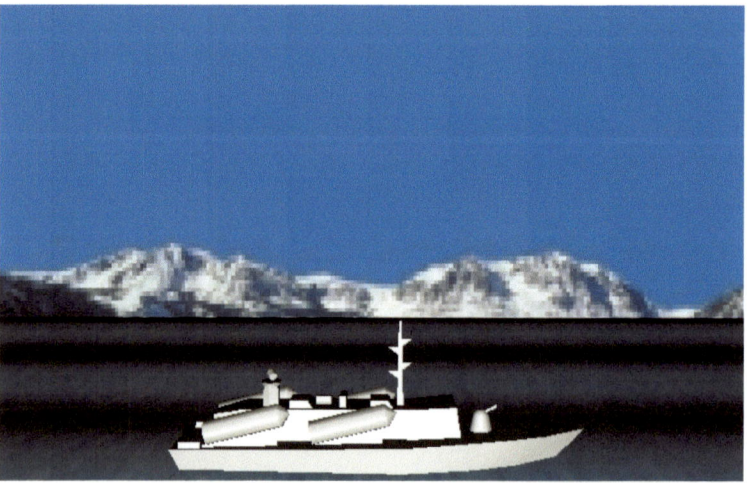

Fig. 7.1 Virtual scene of the sea surface, ship and mountain's background

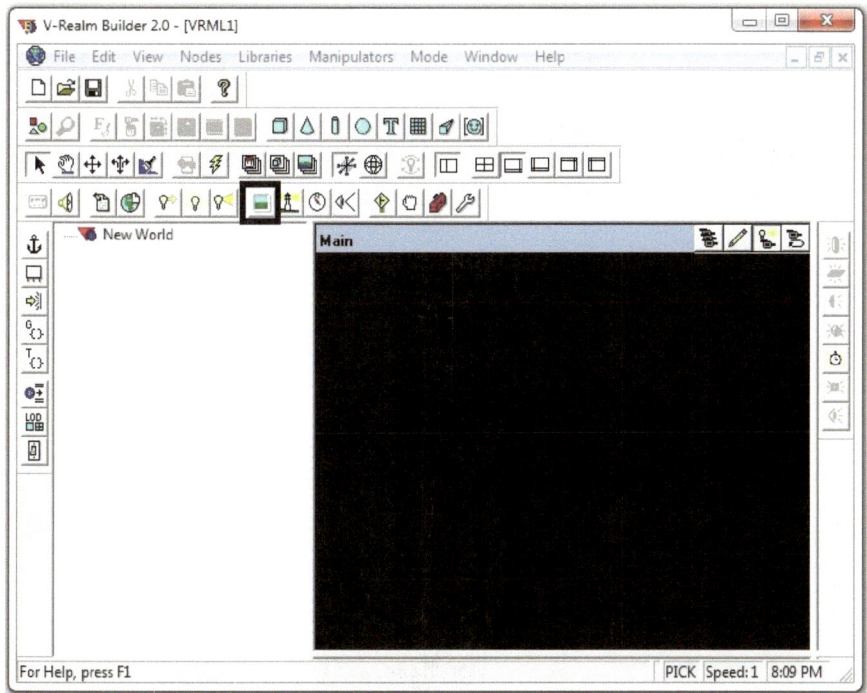

Fig. 7.2 Insert Background button is shown inside the rectangle

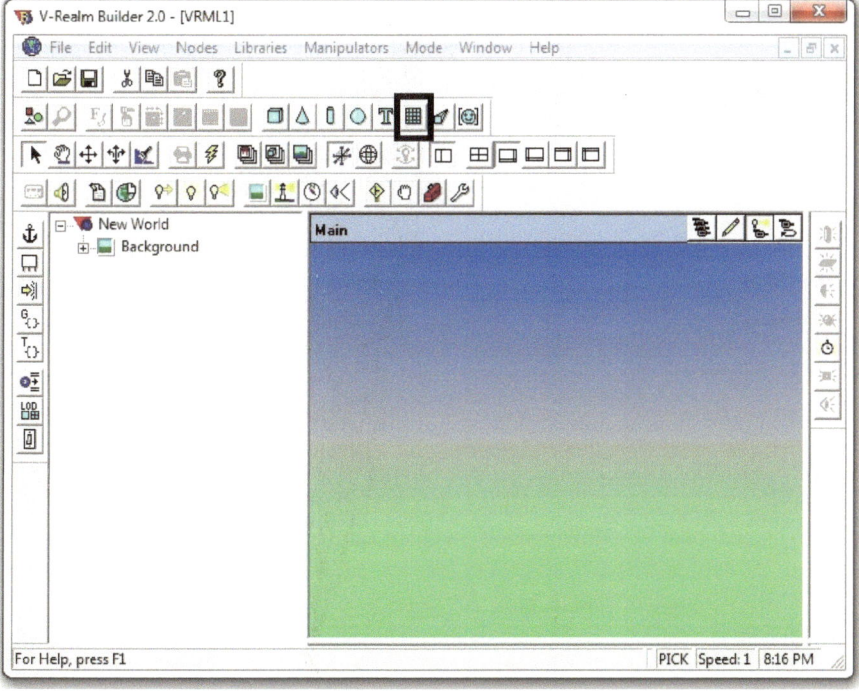

Fig. 7.3 Insert Elevation Grip button is shown inside the rectangle

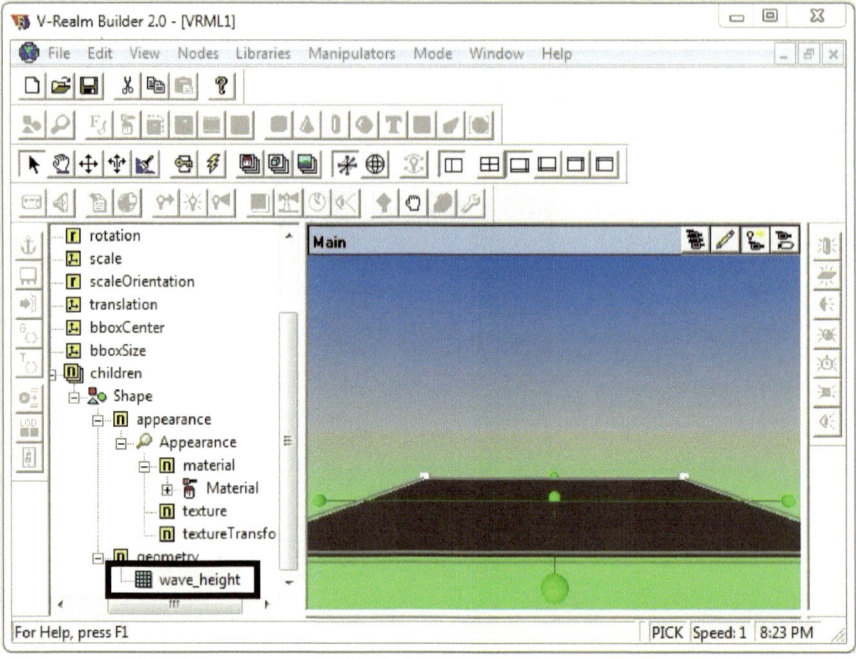

Fig. 7.4   Change the name of the ElevationGrid to wave_height

4. Change the name of **ElevationGrid** to **wave_height** (Fig. 7.4).
5. Double-click **wave_height** to open the **Elevation Grid Editor**. Change the x Dimension, z Dimension, x Spacing, and z Spacing to 31, 20, 20, and 20, respectively (Fig. 7.5). Click OK when done.
6. In following steps, a picture of sea waves will be overlaid on the top of the **Elevation Grid**. Click once on **texture** to select it (Fig. 7.6).
7. Use steps shown in Fig. 7.7 to insert an **Image Texture**.
8. Click on the + sign of **Image Texture** and + sign of **url** to expand them (Fig. 7.8).
9. Double-click on the **S** under **url** to open the **Browse Box** of the **Image Texture** (Fig. 7.9).
10. Download the image *Sea.jpeg* from Chapter 7 material (http://extras.springer.com/) and save it into a local drive. In VRML, click **Browse** to navigate to the directory where you downloaded *Sea.jpeg* (preferably where you want to save your VRML model) and select *Sea.jpeg* image. Press OK when done. Note that VRML will not accept a path for a texture image that has more than 80 characters.
11. Click on the + sign beside **Material** to see the properties underneath it (Fig. 7.10).

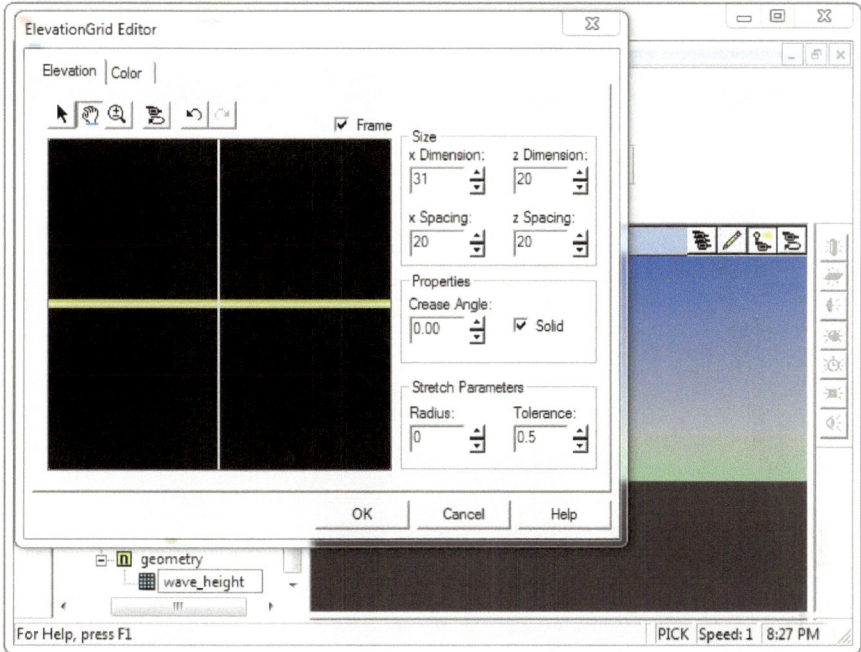

Fig. 7.5 Elevation Grid Editor

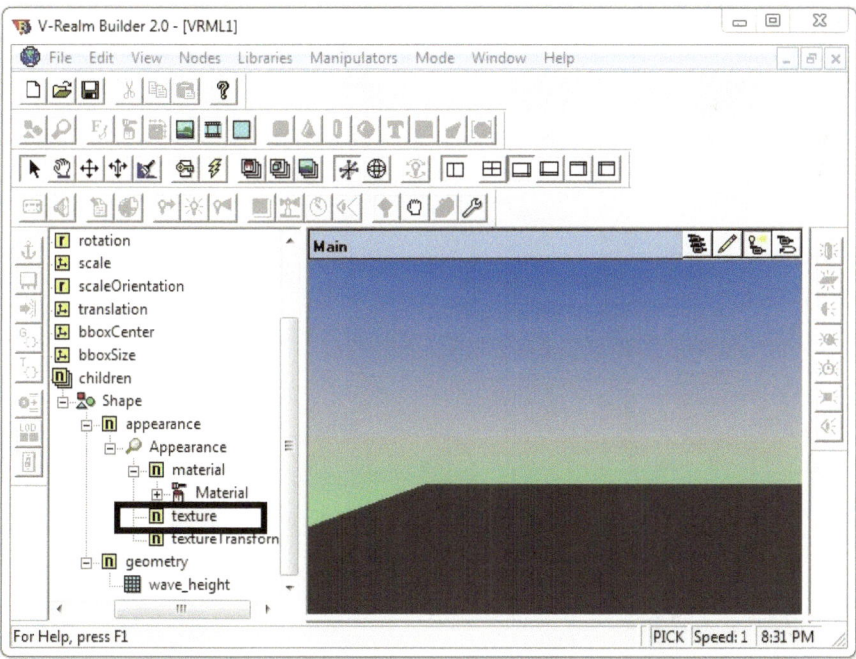

Fig. 7.6 The texture property of the Elevation Grid Transform

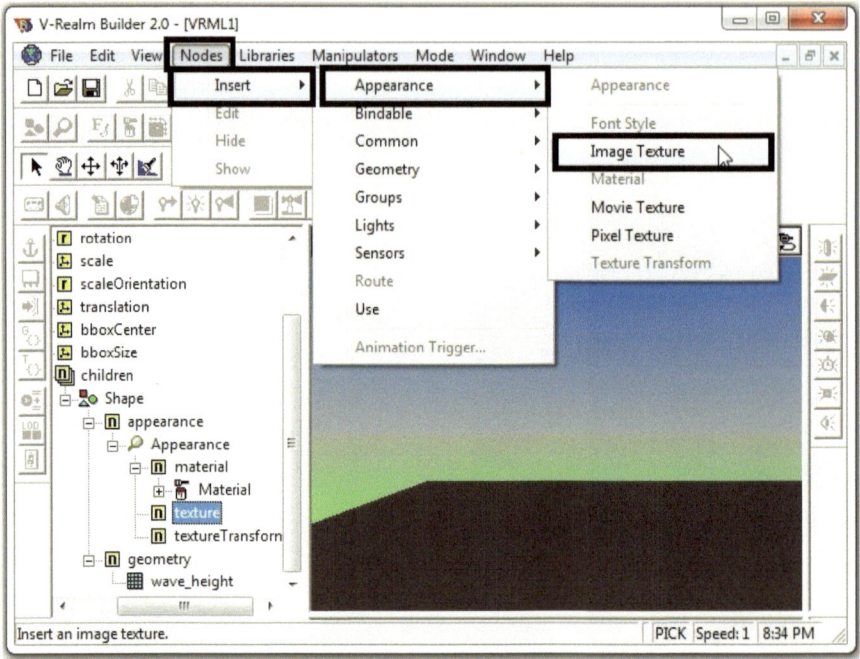

Fig. 7.7 Image Texture button is shown inside the rectangle

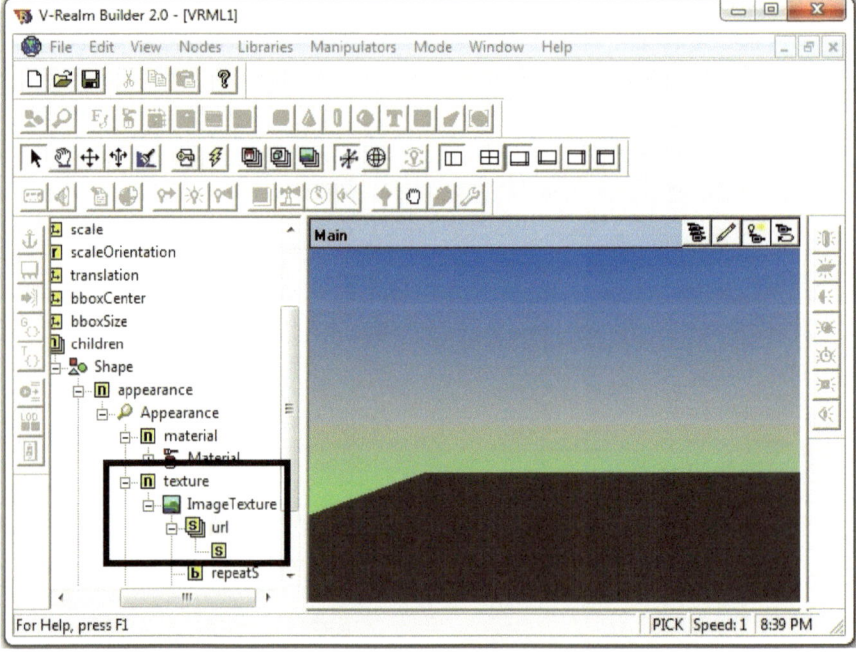

Fig. 7.8 Click on the + sign of Image Texture and + sign of url to expand them

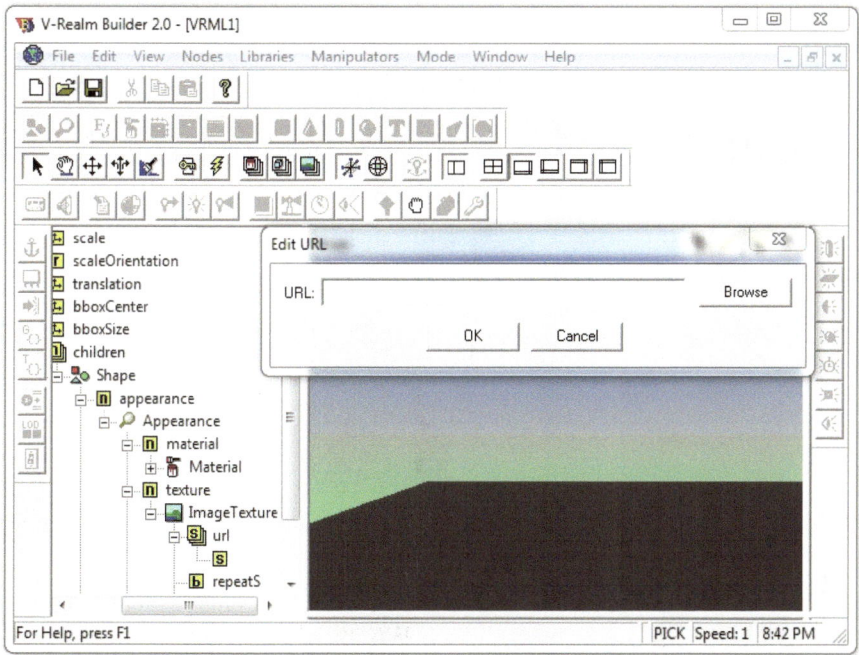

Fig. 7.9 Browse box of the Image Texture

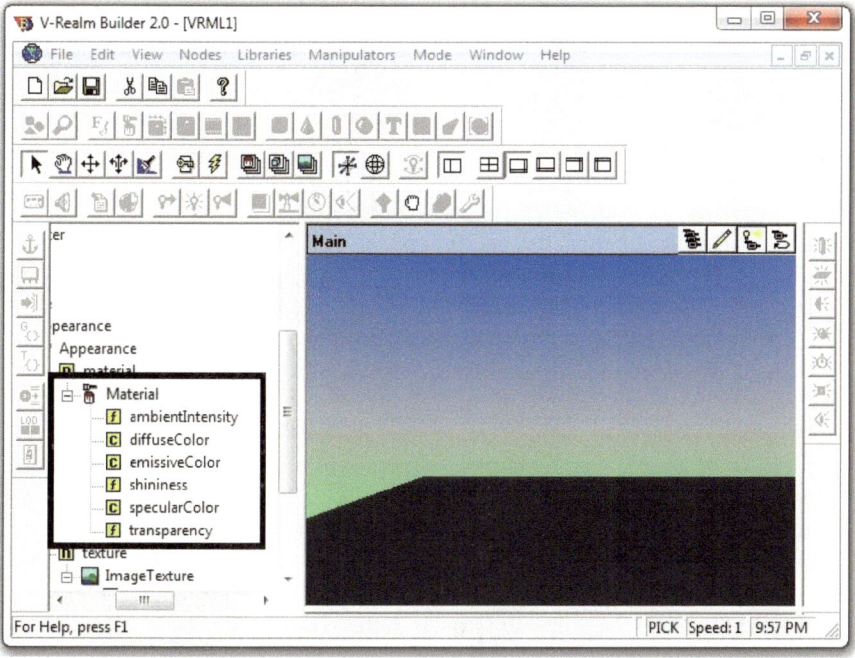

Fig. 7.10 Material properties

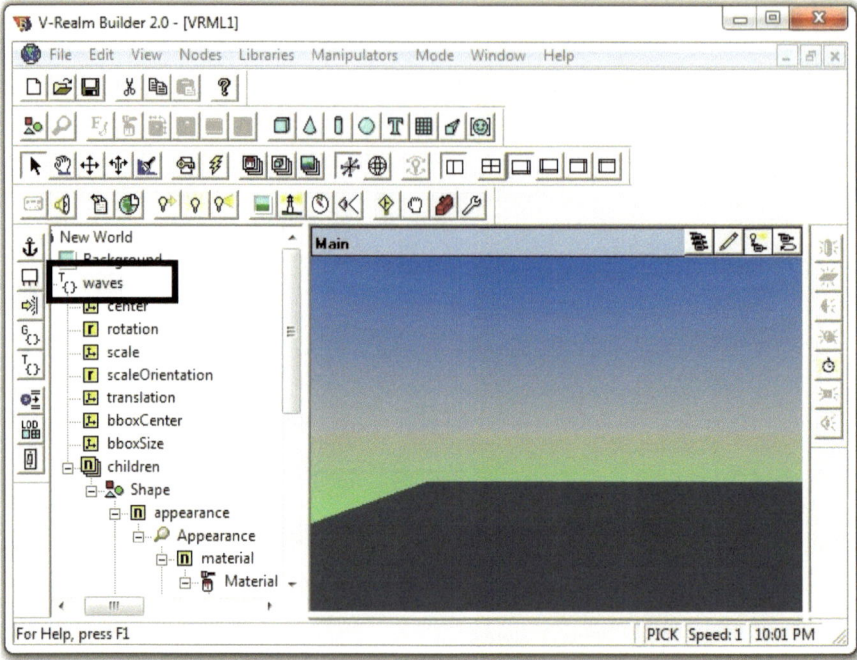

Fig. 7.11 Change the name Transform of the Elevation Grid to waves

12. To make the texture brighter, double-click **ambientIntensity** under **Material** and change the value to 1. Press OK when done.
13. Change the name of the **Transform** of the **Elevation Grid** to **waves** (Fig. 7.11).
14. Double-click the **translation** property of **waves** (Fig. 7.12). Change the x-axis and z-axis values to −295 and −460, respectively. Press OK when done.
15. Click on the top level of the virtual scene (Fig. 7.13).
16. To create a Viewpoint, click **Nodes>Insert>Bindable>Viewpoint** (Fig. 7.14).
17. Click on the + sign on the left of the **Viewpoint** to expand all the subtitles underneath it. Double-click the **position** property of **Viewpoint** (Fig. 7.15) and change the y-axis to 79.
18. To actually observe the scene from the current **Viewpoint** after the changes and check if the position of the current **Viewpoint** is acceptable, double-click the **set_bind** property (Fig. 7.16) and change its value to **True** (this step is not mandatory, and it is only done for checking the **Viewpoint**).
19. In the next steps, a mountain background will be added to the scene. To do so, a wall will be added and placed in the far horizon, and mountain wallpaper will be added to cover the thin wall (similar to the sea surface texture). Click on the top level of the virtual scene (Fig. 7.13).
20. Click on **Insert Box** button (Fig. 7.17) to add the wall of the wallpaper.

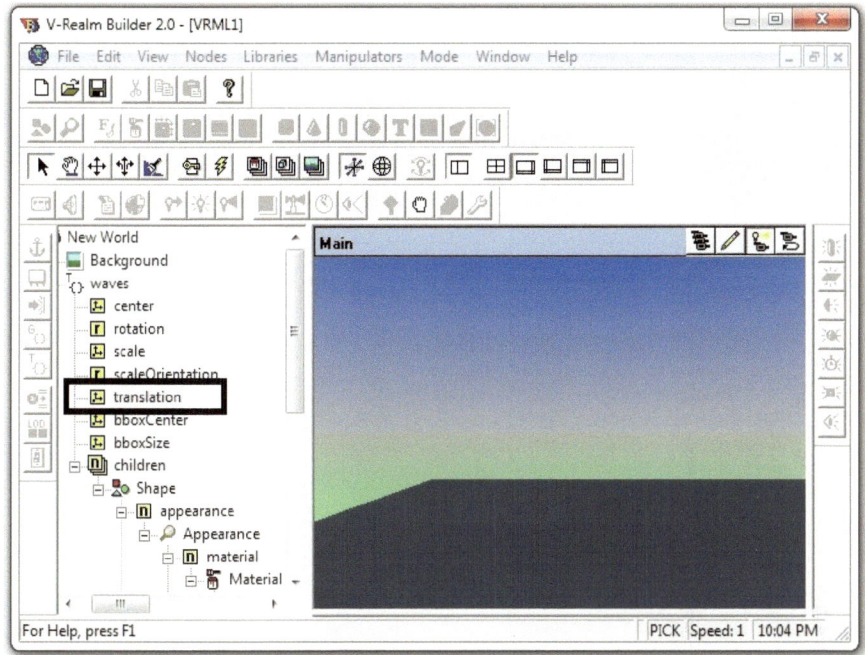

Fig. 7.12 The translation property of waves

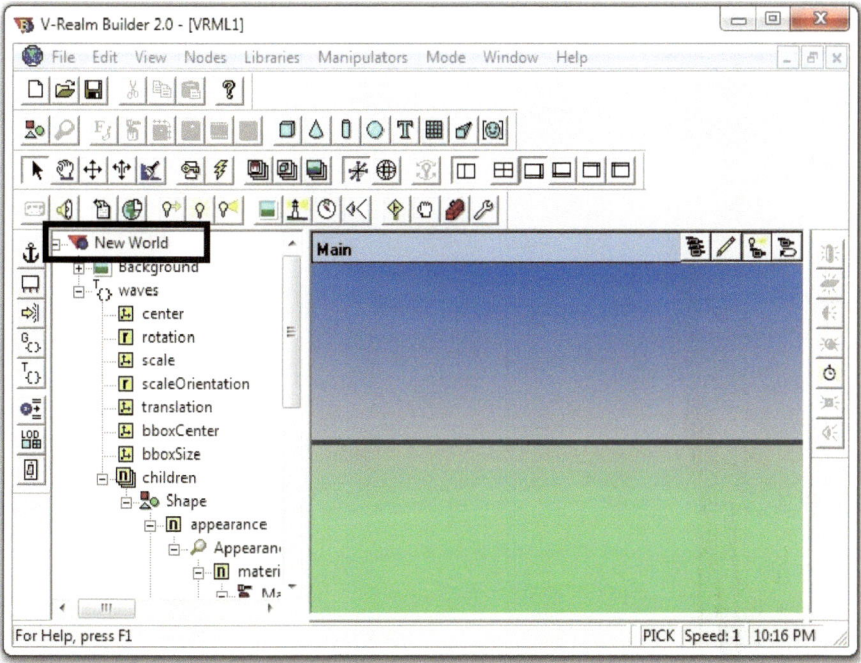

Fig. 7.13 Top level of the virtual scene

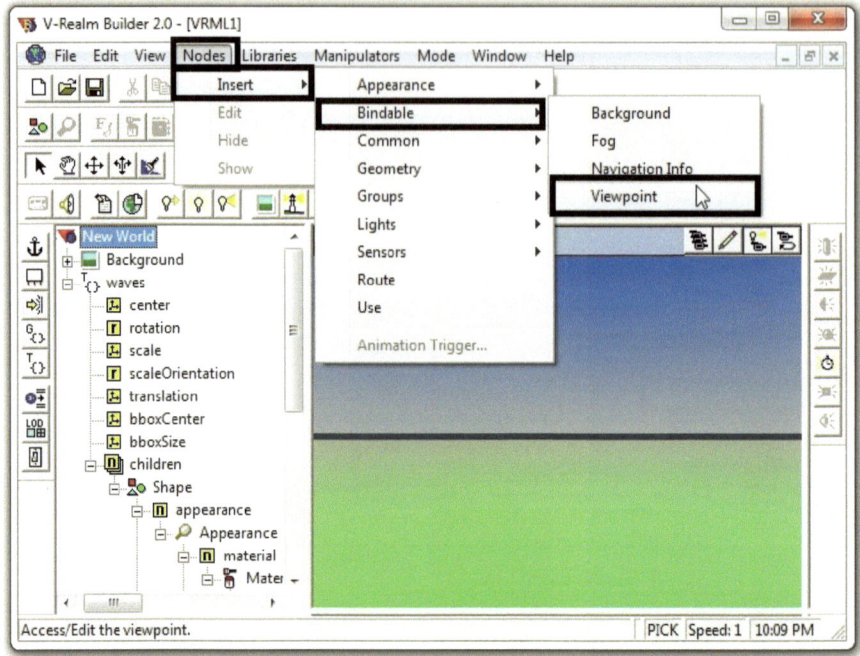

Fig. 7.14 How to create a Viewpoint

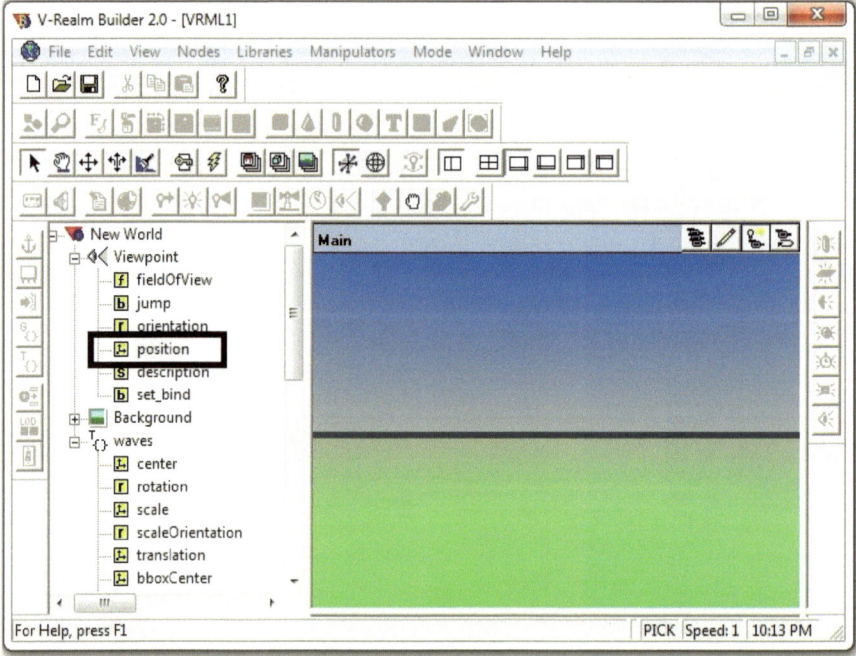

Fig. 7.15 The position property of Viewpoint

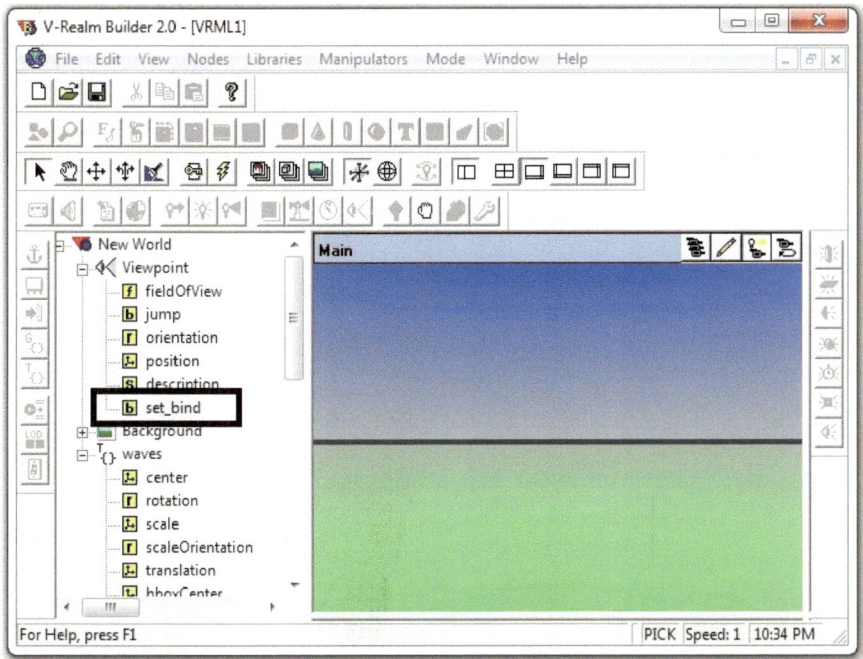

Fig. 7.16 The set_bind property of Viewpoint

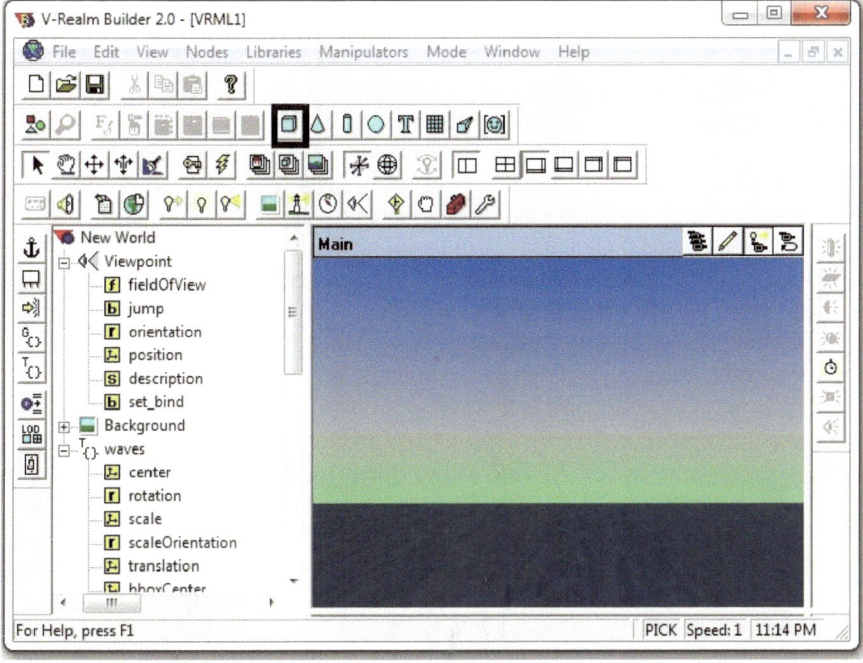

Fig. 7.17 Insert Box button is shown inside the rectangle

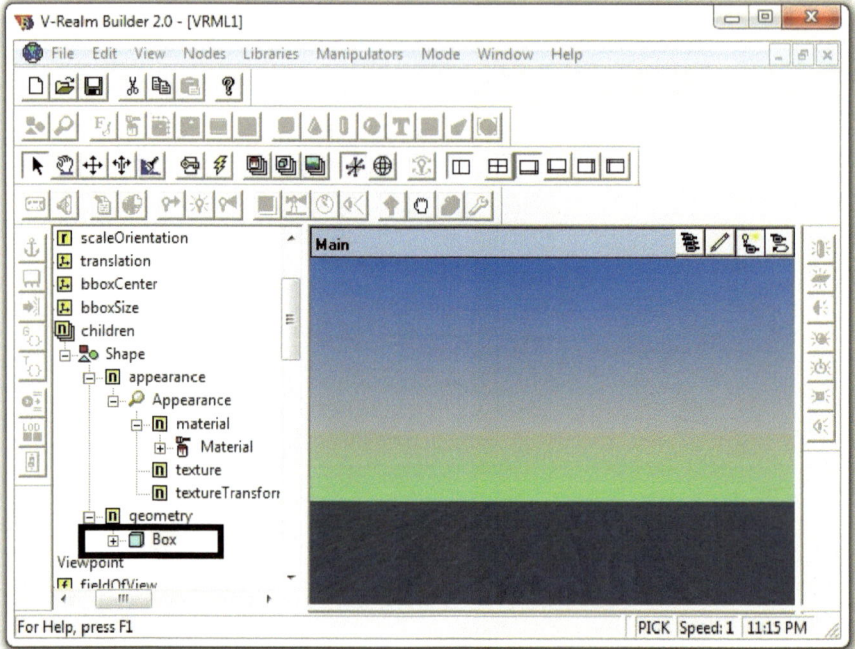

Fig. 7.18 The + sign on the left of Box

21. Click on the + sign to the left of **Box** (Fig. 7.18).
22. Double-click the size property of the **Box**. Change the x-axis, y-axis, and z-axis values to 700, 388, and 2, respectively. Press OK when done. The dimensions of **Box** have been chosen to make sure that the wall paper, when situated far enough at the horizon, will still fill the horizon plane. The user can choose the size of **Box** based on how far it will be placed in the horizon.
23. Double-click the translation property of the **Box** (Fig. 7.19). Change the y-axis and z-axis values to 130 and −500, respectively. Press OK when done. This will translate the **Box** far into the horizon line.
24. Change the name of the **Transform** of the **Box** to **Mountains** (Fig. 7.20).
25. Similar to steps 6–10, a texture image is going to be added for the **Box**. Click once on **texture** property of the **Box** to select it (Fig. 7.21).
26. Use steps shown in Fig. 7.7 to insert an **Image Texture**.
27. Click on the + sign of **Image Texture** and + sign of **url** to expand them (Fig. 7.8).
28. Double-click on the **S** under **url** to open the **Browse Box** of the **Image Texture** (Fig. 7.9).
29. Download the image **Mountain.jpeg** from Chap. 7 material (http://extras. springer.com/) and save it into a local drive. In VRML, click **Browse** to navigate to the directory where you downloaded **Mountain.jpeg** (preferably where

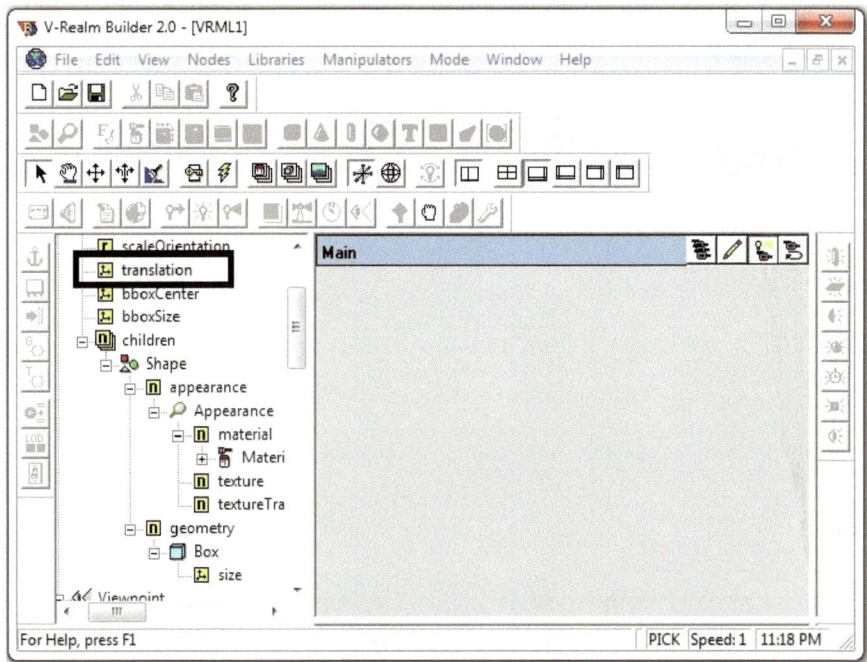

Fig. 7.19 The translation property of Box

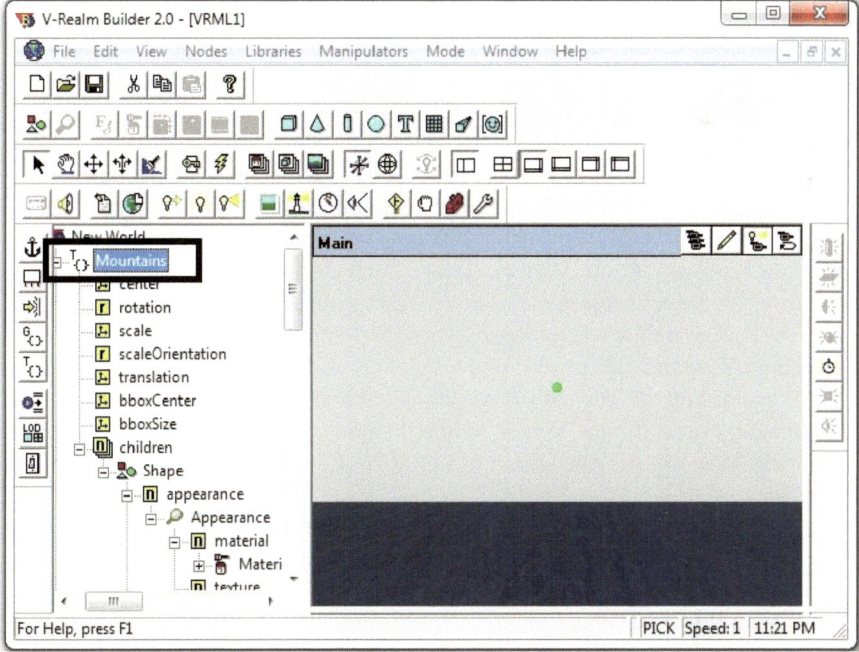

Fig. 7.20 Change the name of the Transform of the Box to Mountains

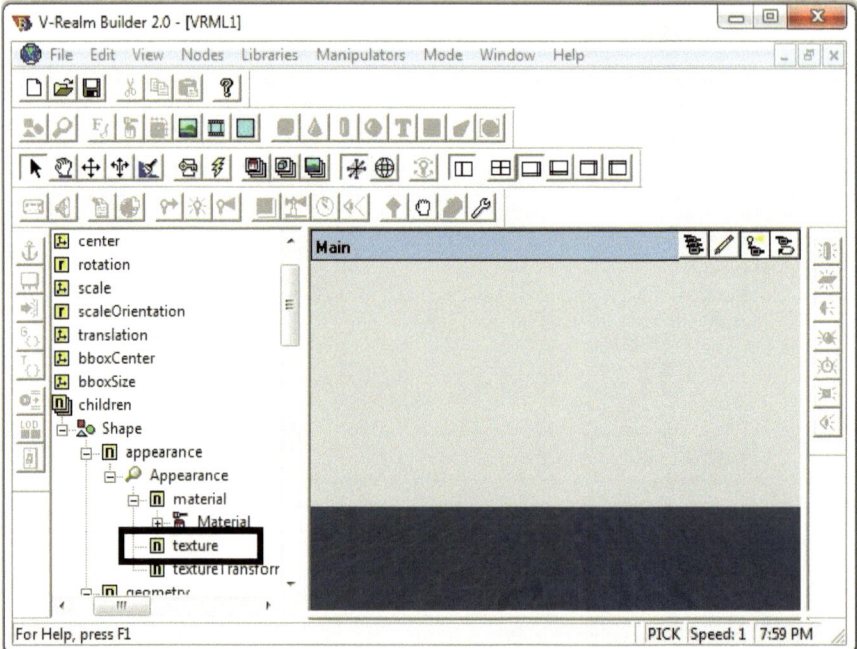

Fig. 7.21 The texture property of the Box

you want to save your VRML model) and select ***Mountain.jpeg*** image. Press OK when done. Note that VRML will not accept a path for a texture image that has more than 80 characters.

30. Click on the top level of the virtual scene (similar to Fig. 7.13).
31. To add a ship object from the object library (which contains many other useful objects), click on **Libraries>Import From…>Object Library** (Fig. 7.22).
32. From the **Object Library**, choose **Transportation (Water)** (Fig. 7.23).
33. From **Transportation (Water)**, choose **Torpedo Boat** (Fig. 7.24).
34. Move the **Select Object** menu to the right to uncover the components of the virtual scene on the left (compare positions 1 and 2 of the **Select Object** Menu in Figs. 7.24 and 7.25).
35. Drag and drop the **Torpedo Boat** on the top level of the virtual scene which is named by default **New World**. While dragging the **Torpedo Boat**, the cursor should look like the one shown in Fig. 7.26. Close **Select Object** menu when done.
36. Click once on the created **Transform** of the **Torpedo Boat** to select it. Click another time on **Transform** and change the name to a meaningful one such as **Ship**. Press Enter when done.
37. Click on the top level of the virtual scene (Fig. 7.13).
38. To add more light into the virtual scene, click **Insert Spot Light** button (Fig. 7.27). The **Spot Light** should be inserted just right after the top level to

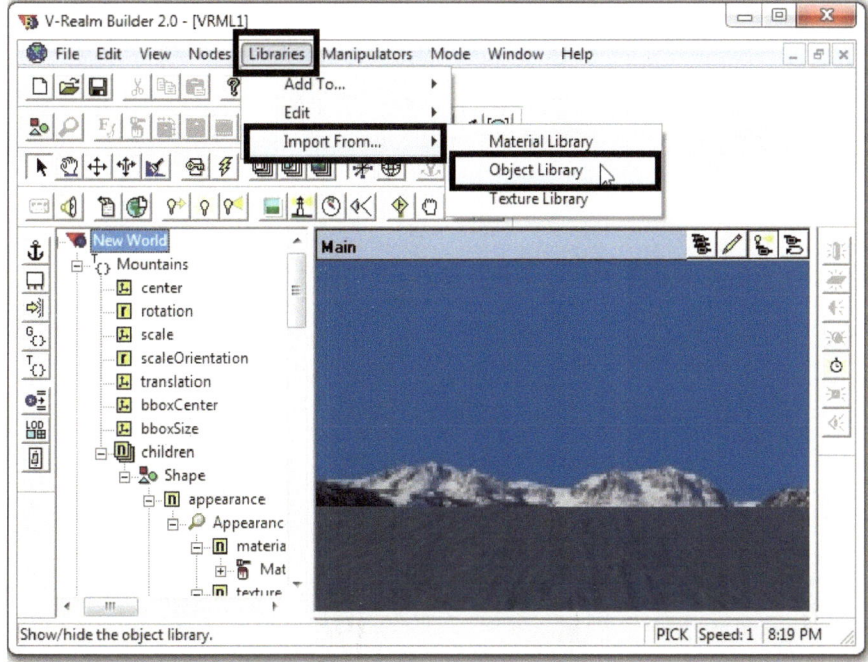

Fig. 7.22 Object library

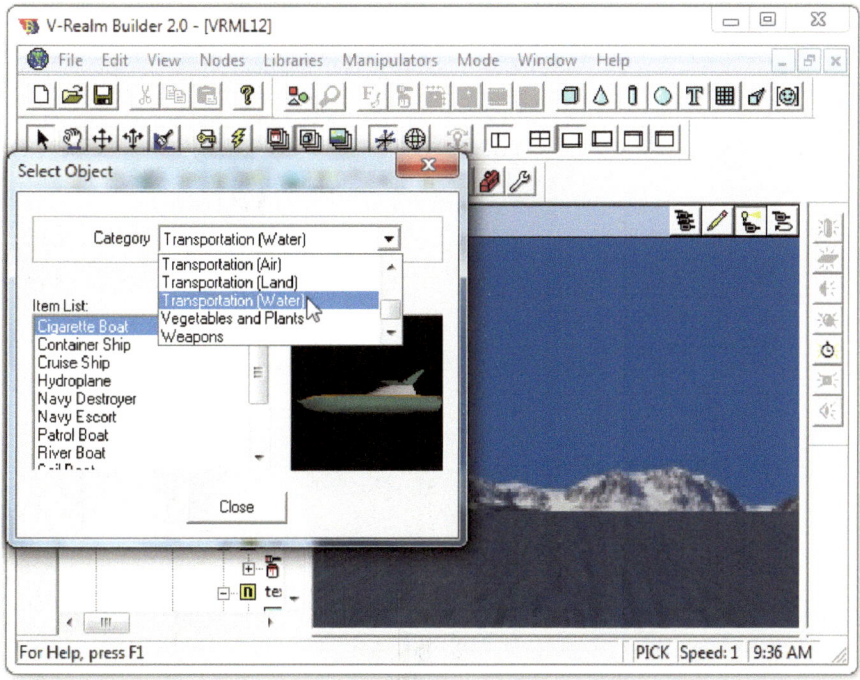

Fig. 7.23 Transportation (Water)

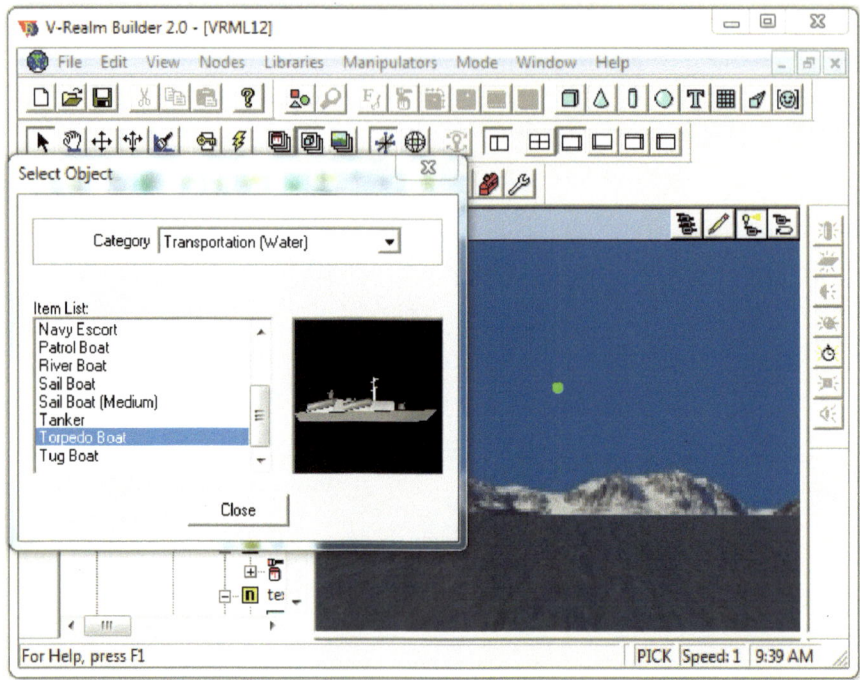

Fig. 7.24 Torpedo Boat

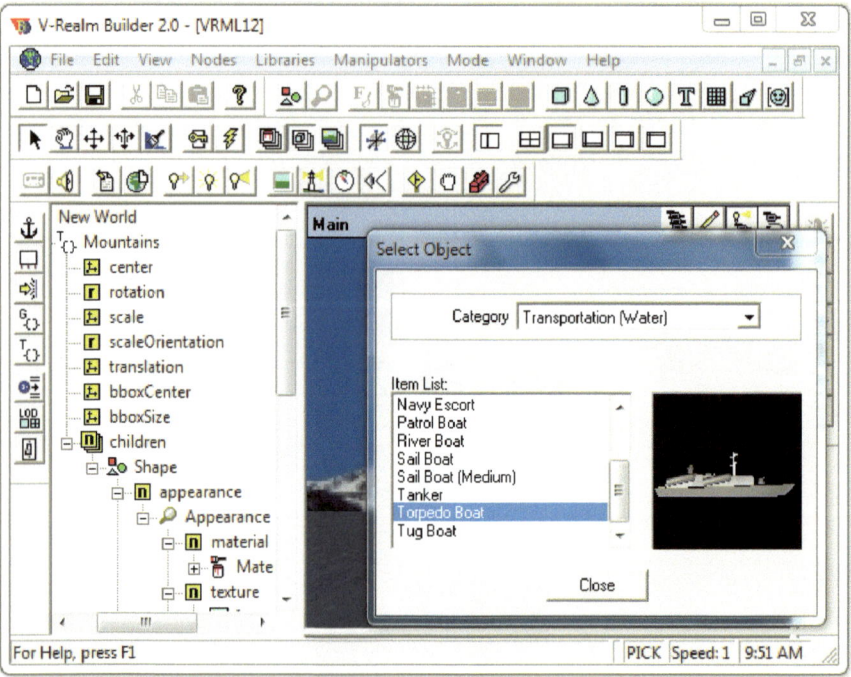

Fig. 7.25 Right position of the Select Object Menu

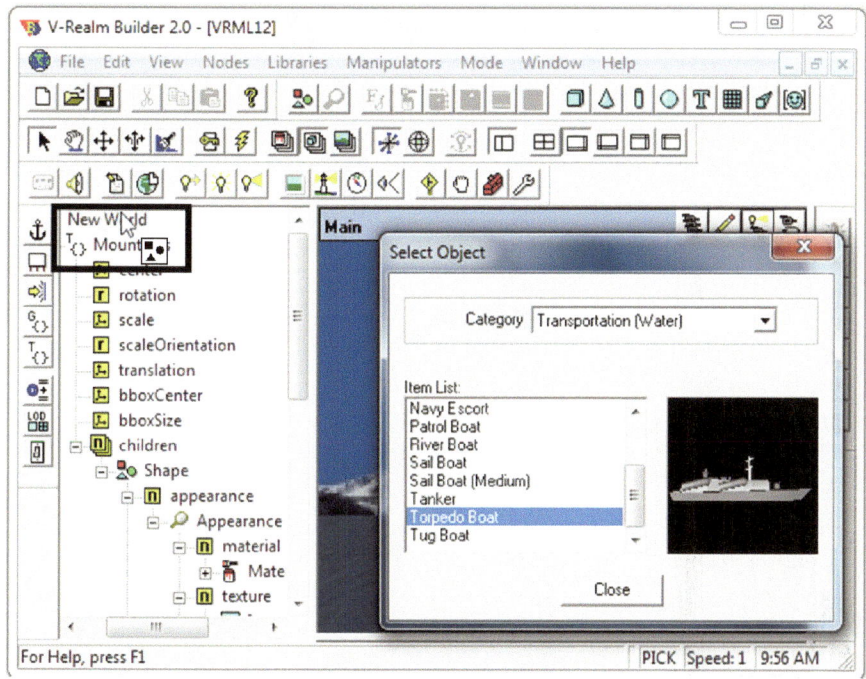

Fig. 7.26 The cursor with the square, triangle, and circle shapes

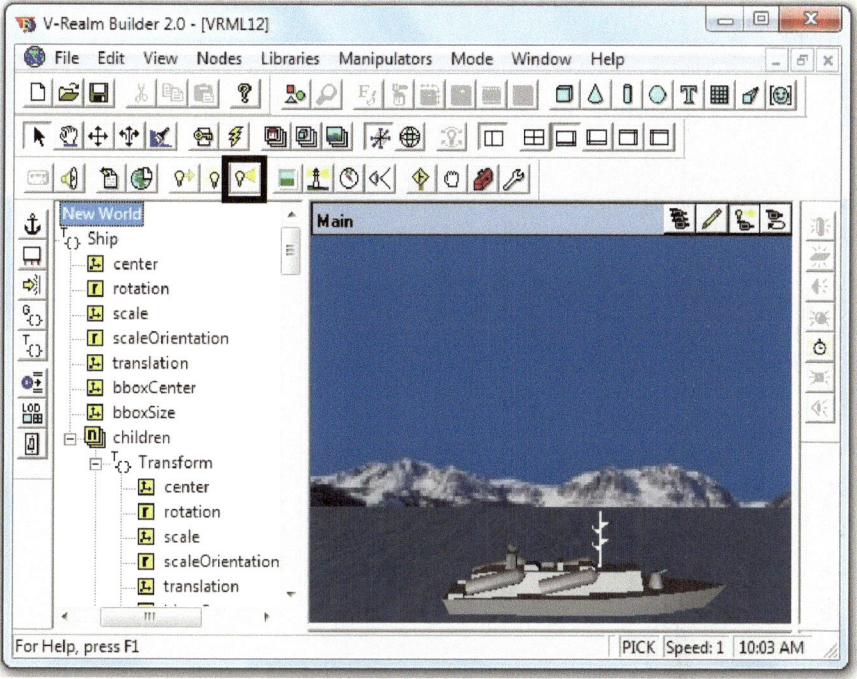

Fig. 7.27 Insert Spot Light button is shown inside the rectangle

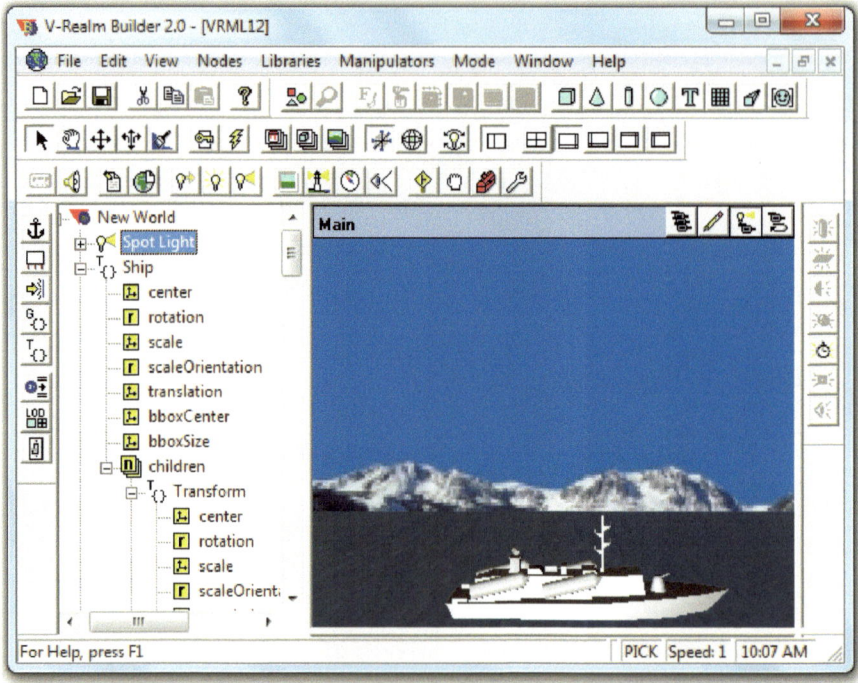

Fig. 7.28 The+sign on the left of Spot Light

illuminate all the rest of the VRML components. It is a good practice to select the top level and add the **Spot Light** after the creation of all the other components.

39. Click on the+sign on the left of the created **Spot Light** (Fig. 7.28).
40. The **Spot Light** will cover more areas of the virtual scene if positioned higher. Double-click the **location property** of the **Spot Light** (Fig. 7.29). Change the *y*-axis value to 52.
41. Save the virtual scene when done in the working directory of MATLAB®: **File>Save As>***sea_scene*.

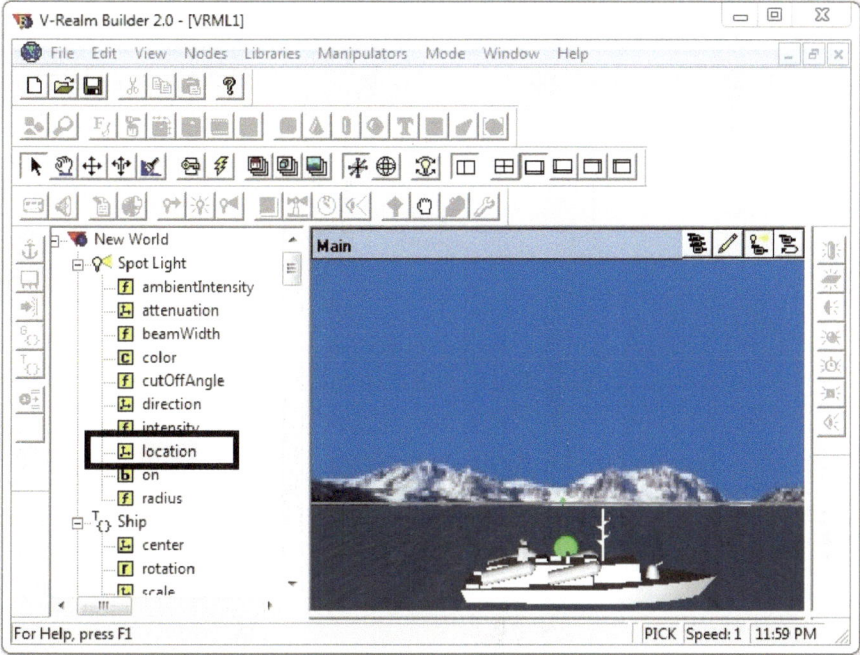

Fig. 7.29   The location property of Spot Light

7.4   M-script File for Animating the Virtual Scene

In what follows, the M-script file that will animate the virtual scene, *sea_scene.wrl*, will be presented. The sea surface will vary as a sinusoidal function of time. The ship will move along the *x*-axis of the virtual scene. The variables that will be changed in the virtual world are:

1. The **height** property of **wave_height** based on Eq. 7.1
2. The **translation** property of **Ship** based on Eq. 7.2

The following is the M-script that is used to animate the virtual model of the ship and waves. The script *sea_animate.m,* the virtual scene *sea_scene.wrl*, the texture images, and a video recording for the animation can be downloaded from Springer's web site http://extras.springer.com/. The files can be found in folder Chapter 7.

sea_animate.m

```
%%%%%%%%%%%%%%%%%%%%%%%%%%%%%%%%%%%%%%%%%%%%%
%  initialize the virtual model %
%%%%%%%%%%%%%%%%%%%%%%%%%%%%%%%%%%%%%%%%%%%%%
% create a virtual world associated with sea_scene.wrl file
world = vrworld('sea_scene.wrl');
% open the virtual world
open(world);
% draw the virtual world
fig = view(world, '-internal');
%update the virtual scene
vrdrawnow;


%%%%%%%%%%%%%%%%%%%%%%%%%%%%%%%%%%%%%%%%%%%%%%
%define the size of the Elevation Grid of the wave(dimensions taken fro
the virtual scene sea_scene.wrl)
%%%%%%%%%%%%%%%%%%%%%%%%%%%%%%%%%%%%%%%%%%%%%
xSpacing=20;
zSpacing=20;
xDimension=31;
zDimension=20;


%%%%%%%%%%%%%%%%%%%%%%%%%%%%%%%%%%%%%%%%%%%%%
%construct a 2D mesh for the Elevation Grid of the wave
%%%%%%%%%%%%%%%%%%%%%%%%%%%%%%%%%%%%%%%%%%%%%
X=0:xSpacing:(xDimension-1).*xSpacing;
Z=0:zSpacing:(zDimension-1).*zSpacing;
Mesh_zx=meshgrid(Z,X);


%%%%%%%%%%%%%%%%%%%%%%%%%%%%%%%%%%%%%%%%%%%%%
%%% for loop to run the script  %%%
%  across the entire desired time%%
%%%%%%%%%%%%%%%%%%%%%%%%%%%%%%%%%%%%%%%%%%%%%
for t=0:0.1:40
    %change the translation property of Ship
    world.Ship.translation=[-100+t*10 0 0];

    %compute the sinusoidal time varying height of waves
    height=2.*sin(3.*t+Mesh_zx);

    %change the height vector of dimensions (xDimension,zDimension)
    %to a  vector of dimensions (xDimension*zDimension,1) as VRML
    % expects the height property of Elevation Grid to be column vector.
    height_reshape=reshape(height,xDimension*zDimension,1);

    %change the height property of wave_height
```

```
world.wave_height.height=height_reshape;

%update the virtual scene
vrdrawnow;

%set the time of the virtual world to be t
set(world,'Time', t);

%delay the execution of the next time step for 0.05 seconds to
%make the execution time closer to real time
pause(0.05);
end
```

7.5   Application Problem: Ship Heading Controller

The main objective of this section is to design a controller for the ship heading ψ (the angle of the ship with the x-axis in Fig. 7.30) by commanding the rudder angle α. A virtual scene for the ship (similar to *sea_scene.wrl*) will be constructed to implement the controller. The proposed controller for this problem is the fuzzy logic controller. The user can choose any control technique to control the heading. The fuzzy controller has two inputs and one output. The inputs are the normalized error and its time derivative, whereas the output is the normalized control action which yields the angle of the rudder, α, when multiplied by the gain K_α (Fig. 7.31). The

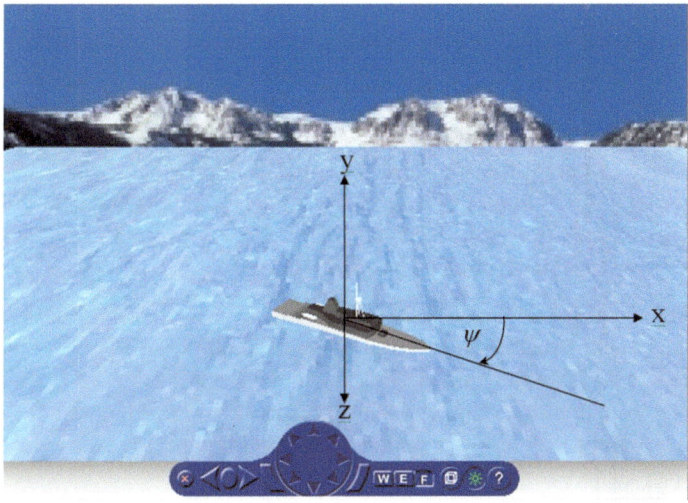

Fig. 7.30 Heading of the ship

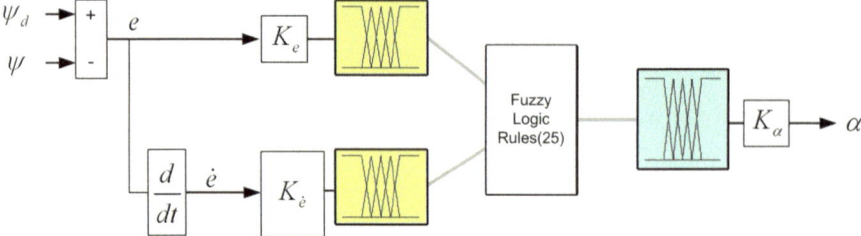

Fig. 7.31 Fuzzy logic heading controller

simplified relation between ship heading ψ and the rudder angle α is given by (for a more detailed equation, refer to Khaled 2010)

$$\ddot{\psi} = \frac{M + \sin(\alpha)}{10}, \tag{7.3}$$

where M is the sum of the external moments around the y-axis of the ship. For this example, $M = 0.5\sin(3t)$ and $\alpha \in [\frac{-\pi}{5}, \frac{\pi}{5}]$. Note that rudder machines are practically ineffective to control the heading of the ship at low velocities of ships, but to simplify the problem, it was assumed otherwise.

Solve for the following:

1. Reconstruct the virtual scene shown in Fig. 7.30 by repeating the same steps outlined in Sect. 7.3. Save the virtual scene as *sea_scene2.wrl*. The following are the differences from Sect. 7.3:

   - In step 17, set the **position** property of **Viewpoint** in the y-axis and z-axis to 100 and 25, respectively. Also set the **orientation** property to −1, 0, 0, and 20 in the x-axis, y-axis, z-axis, and Rotation boxes, respectively. This will allow the viewer to look into the virtual scene downward from an angle of 20° along the z-direction.
   - In step 33, add a **Patrol Boat** instead of a **Torpedo Boat**. In the **translation** property of **Patrol Boat**, change the z-axis to −150.
   - After adding the **Spot Light** (step 40), change the **cutOffAngle** property of the **Spot Light** to 1.57 to allow for more space to be illuminated. In addition, change the **ambientIntensity** property to 1.

2. Using the MATLAB® Fuzzy Logic Toolbox (type in **fuzzy** in the MATLAB® workspace), construct the fuzzy controller based on the rules given in Table 7.1 and the membership functions in Figs. 7.32, 7.33, and 7.34.

3. Using the equation of motion of the heading given in Eq. 7.3, obtain the gains for the fuzzy controller (K_e, $K_{\dot{e}}$, and K_α) that will control the ship's heading to a desired heading $\psi_d = \frac{\pi}{3}$ with a maximum steady state error of 0.01 rad starting from zero initial conditions ($\psi = 0$ and $\dot{\psi} = 0$).

Table 7.1 Rules for the heading fuzzy logic controller

$\dot{e}$ \ e	NL	NS	Z	PS	PL
NL	NL	NL	NL	NL	PL
NS	NL	NS	NS	Z	PL
Z	NL	NS	Z	PS	PL
PS	NL	Z	PS	PS	PL
PL	NL	PL	PL	PL	PL

NL stands for negative large, *NS* stands for negative small, *Z* stands for zero, *PS* stands for positive small and *PL* stands for positive large

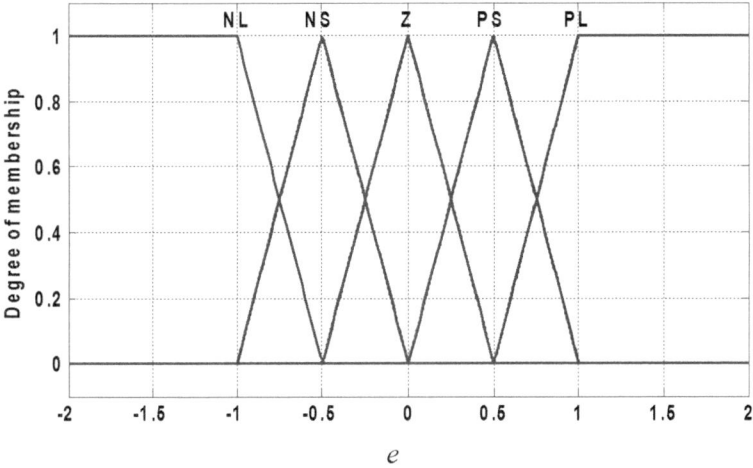

Fig. 7.32 Membership functions for the error *e*

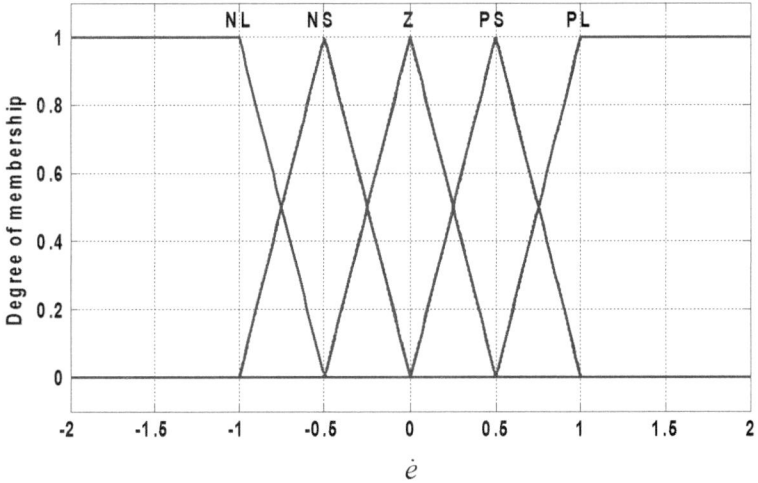

Fig. 7.33 Membership functions for the time derivative of the error $\dot{e}$

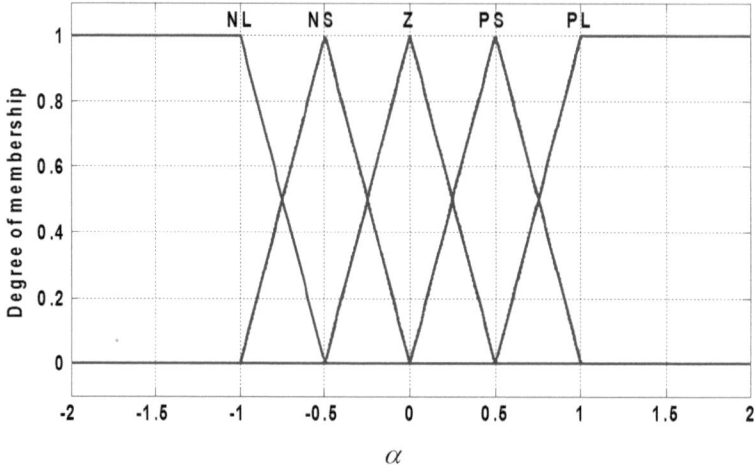

Fig. 7.34 Membership functions for the normalized controller output

4. Write an M-script file that will:

   - Animate the sea waves of ***sea_scene2.wrl*** based on Eq. 7.1
   - Call the developed fuzzy logic controller for the heading to compute the rudder angle α
   - Compute the heading of the ship based on the rudder angle provided by the controller and using the equation of motion of the heading given by Eq. 7.3
   - Change **rotation** property of **Patrol Boat** based on the angle ψ. (Hint: rotate the **Patrol Boat** around the y-axis by an angle ψ.)

The solution of the problem can be downloaded from Springer's web site http://extras.springer.com/. The files can be found in folder /Chapter 7/Application Problem.

References

Books

Khaled N (2010) Guidance and control of ships: modeling, estimation, guidance and control. VDM Verlag Dr. Müller, Saarbrücken

Perez T (2005) Ship motion control: course keeping and roll stabilisation using rudder and fins. Springer, New York

Chapter 8
Animation of a Translating Cube Using Simulink®

8.1 Introduction

The purpose of this chapter is to introduce Simulink® users that are interested in animating their physical system in a virtual reality environment to the virtual reality tool, V-Realm Builder (VRML). This chapter will walk the new user through a complete exercise on how to animate a physical problem governed by an equation(s) of motion in VRML environment using Simulink®.

In Sect. 8.2, a simple dynamics problem of a cube that is subjected to an external force will be presented. Newton's second law will be used to derive the equation of motion.

The Cartesian coordinate convention of VRML will be briefly explained in Sect. 8.3.

To animate the virtual scene, a Simulink® model will be developed in Sect. 8.4. Furthermore, a procedure to save the animation as a movie file will be included.

The chapter concludes by an application problem of two colliding masses in Sect. 8.5.

The electronic version of all the Simulink® models and VRML files in addition to a video recording for the animation of the cube can be downloaded from Springer's web site http://extras.springer.com/.

8.2 Translating Cube Problem

Given a cube of mass $m = 1$ (kg) and of dimensions $0.5 \times 0.5 \times 0.5$ (m^3). The cube is subjected to the following external force: $F_z = -1$ (N).

Matlab® and Simulink® are registered trademark of The Mathworks, Inc.

N. Khaled, *Virtual Reality and Animation for MATLAB® and Simulink® Users: Visualization of Dynamic Models and Control Simulations*, DOI 10.1007/978-1-4471-2330-9_8, © Springer-Verlag London Limited 2012

Using Newton's second law, the following second-order differential equation for the z-displacement of the cube can be developed:

$$m\ddot{z}(t) = F_z \Rightarrow \ddot{z}(t) = \frac{F_z}{m} \qquad (8.1)$$

The purpose of this exercise is to animate the cube based on the solution of this equation for zero initial conditions. The virtual reality scene for the cube was constructed in Chap. 3.

8.3 Cartesian Coordinates for the Virtual Reality Environment VRML

The plane of the computer screen is of two dimensions. In VRML the x-axis is the horizontal line of the screen and points to the right. The y-axis is vertical and points upward. As for the z-axis, it is perpendicular to the plane of the screen, and it is pointing out of the screen toward the user (Fig. 8.1). For further information, the reader is referred to Sect. 3.3.

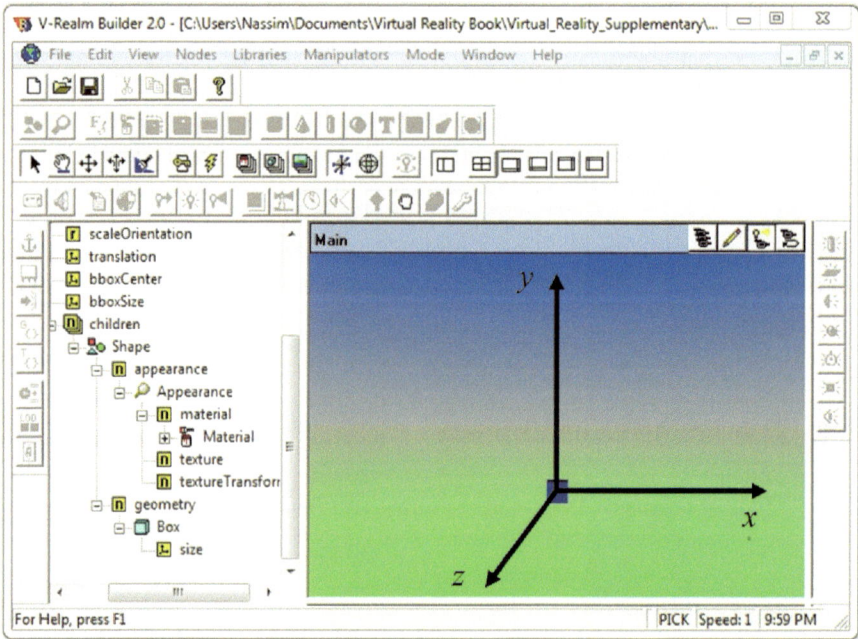

Fig. 8.1 Cartesian coordinates for VRML

8.4   Creating the Simulink® Model for Animating the Virtual Scene

In what follows, the virtual scene of the cube, *Cube_Virtual.wrl*, will be animated using Simulink®. The only element of the virtual scene that will be animated is the cube. Its translation property will be changed based on the solution of Eq. 8.1.

Follow the following steps to create the Simulink® model that will animate the virtual scene (assuming that the user has the **Virtual Reality Toolbox*** installed in Simulink®):

1. Create a new Simulink® model.
2. Add a **VR Sink** from the Virtual Reality Toolbox* to the current model (Fig. 8.2). This sink will contain the virtual scene *Cube_Virtual.wrl* and will allow the user to manipulate the scene through Simulink® inputs.
3. Double-click the **VR Sink** in the model. A window for the parameters of the **VR Sink** will open (Fig. 8.3). Click on the **Browse** button (Fig. 8.3) to navigate and select the virtual scene *Cube_Virtual.wrl* which can be found in Chap. 8

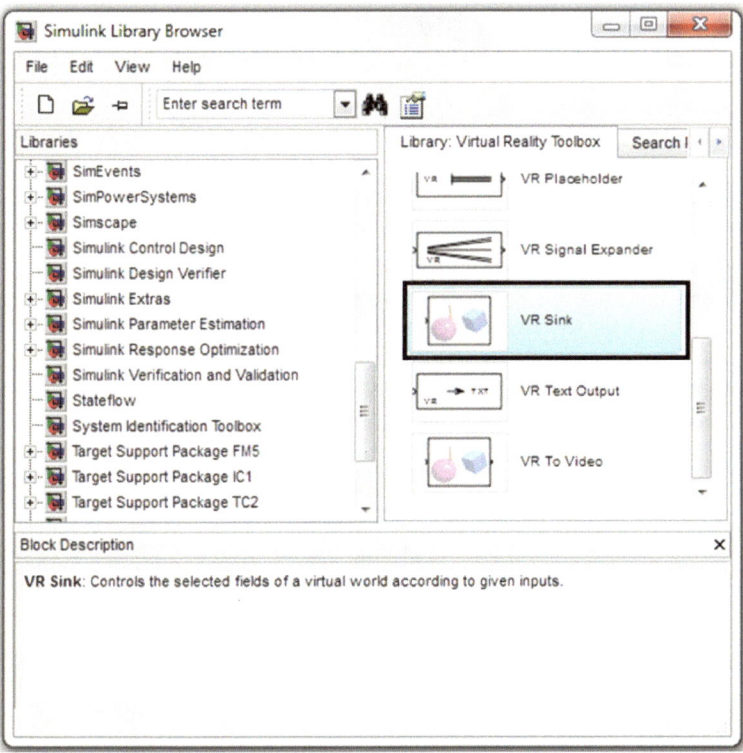

Fig. 8.2   VR Sink in virtual reality toolbox*

---

* As of Matlab® 2009a, **Virtual Reality Toolbox** has been renamed **Simulink 3D Animation**.

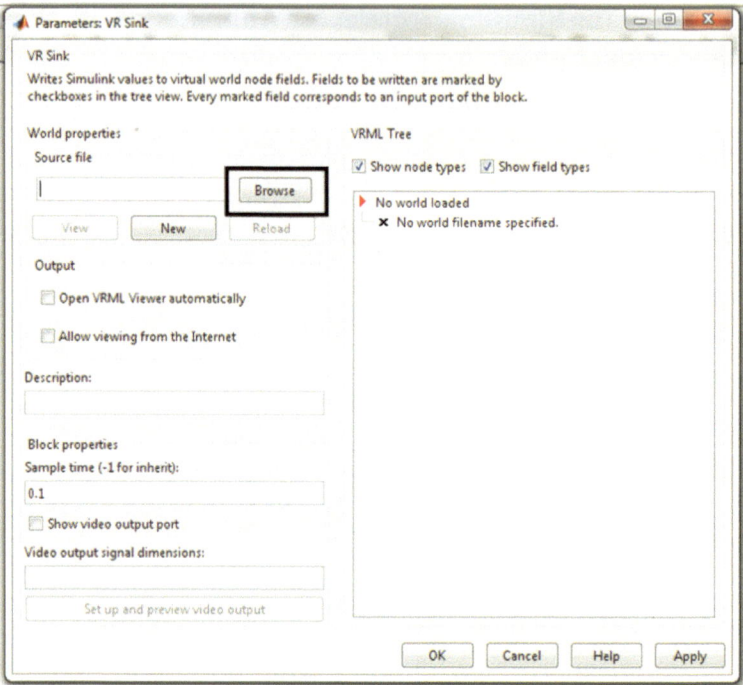

Fig. 8.3 Browse button for the virtual scene is shown in the rectangle

material of the book. Download the file to the working directory of MATLAB®. Click **Open** once you select the file *Cube_Virtual.wrl*.

4. Click on the + sign to the left of **Cube (Transform)** (Fig. 8.4).

5. Tick the square to the left of **translation** property to select it (Fig. 8.5). Click OK to accept the changes and close the parameters of the **VR Sink**.

6. In the Simulink® model, go to **Simulation>Configuration Parameters** and change the solver type to **Fixed-step** (Fig. 8.6). Set the **step size** to 0.1 to match the **Sample time** of the **VR Sink**. Click OK to accept the changes.

7. Figure 8.7 shows the final model that is used to animate the virtual scene. The user has to be careful about the dimension size of the signals going into **VR Sink** for **Cube.translation**. It is a 3×1 vector.

8. To record the simulation into a movie, double-click the **VR Sink** to open the virtual scene (Fig. 8.8).

9. Click on **Recording** and then choose **Capture and Recording Parameters…** to set up the recording options for the animation (Fig. 8.9).

10. To record the animation, tick the square to the left of **Record to AVI** (Fig. 8.10). Click OK when done. The first time the user runs the simulation, the default name of the animation will be saved as *Cube_Virtual_anim_1.avi* (*Cube_ Virtual.wrl* is the name of the virtual scene). Index 1 will be automatically incremented each time the user reruns the simulation; thus, multiple simulations will be saved rather than overwritten.

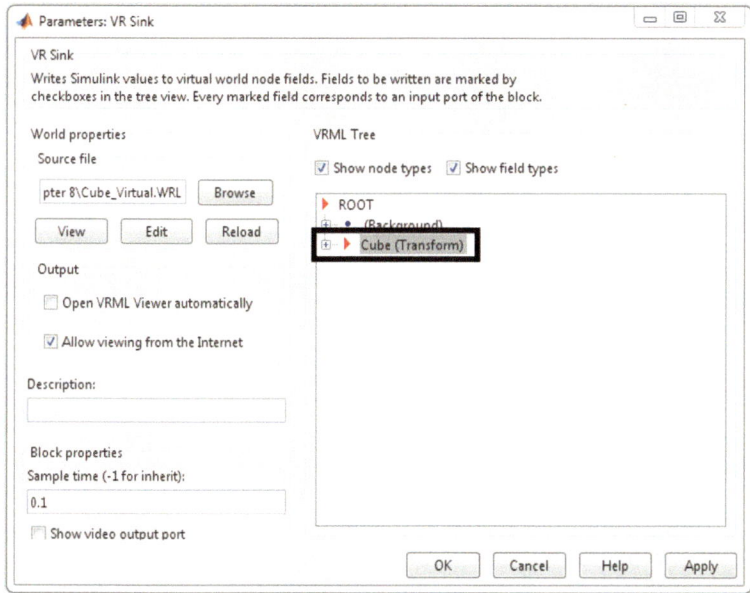

Fig. 8.4 Cube (Transform) is shown in the rectangle

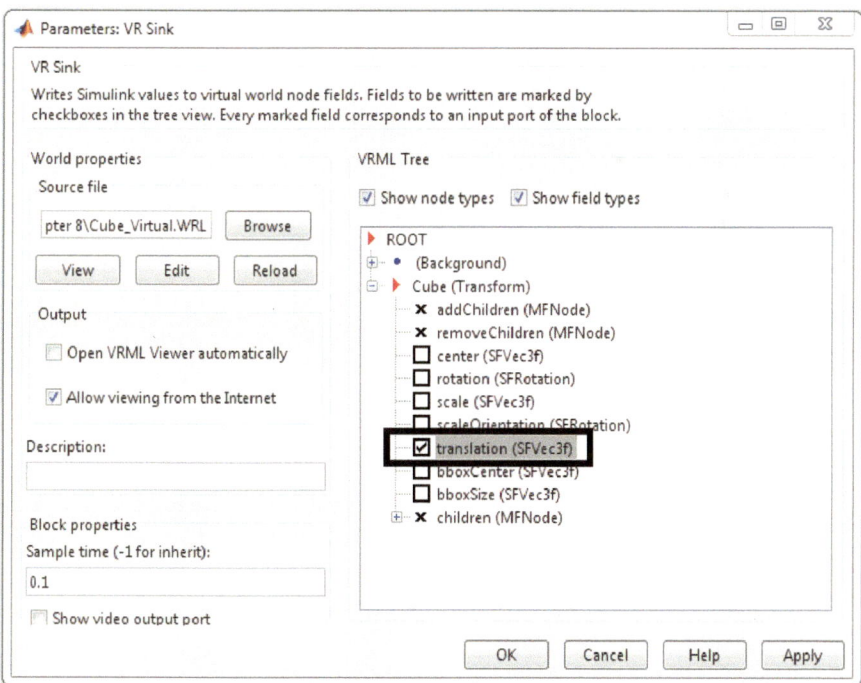

Fig. 8.5 The square to the left of translation property is shown in the rectangle

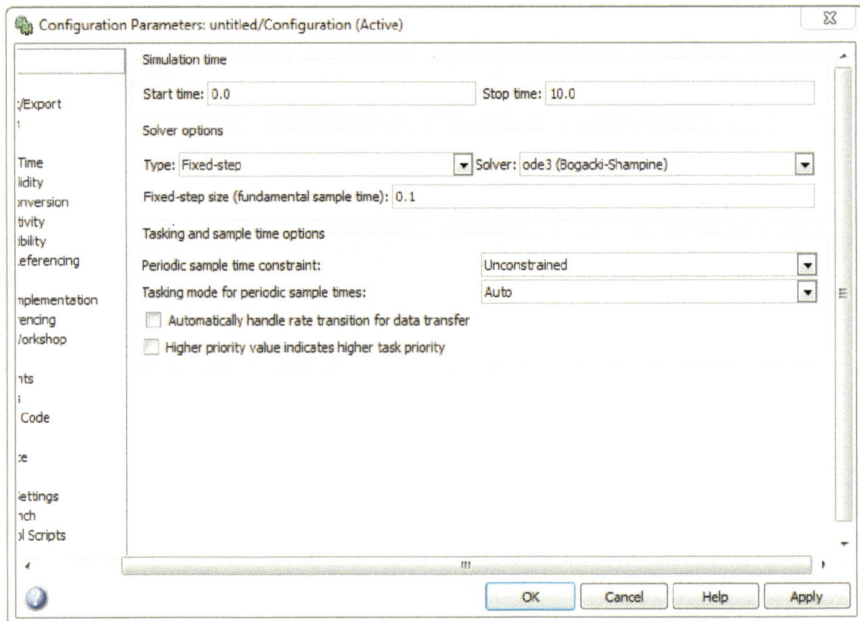

Fig. 8.6 Solver options

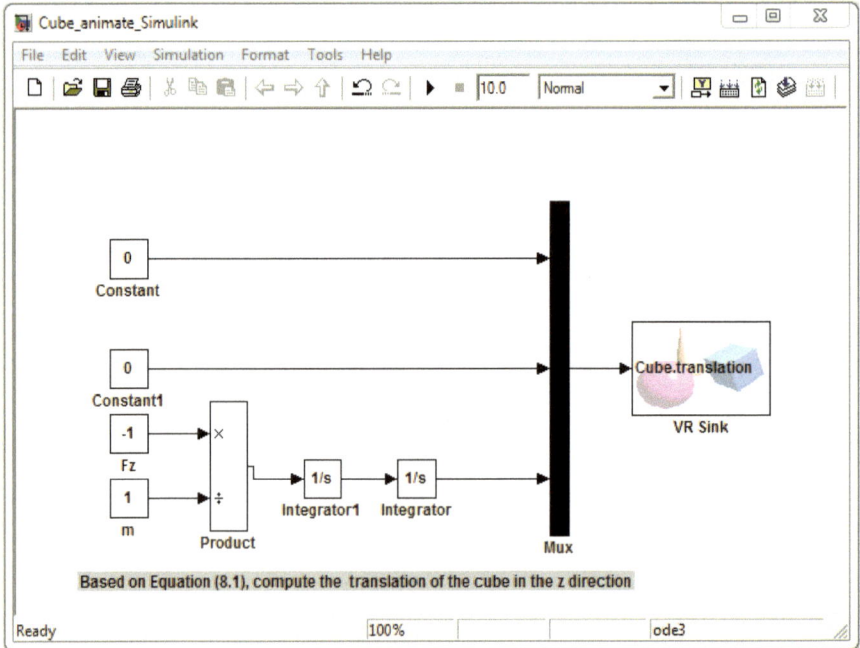

Fig. 8.7 Final model

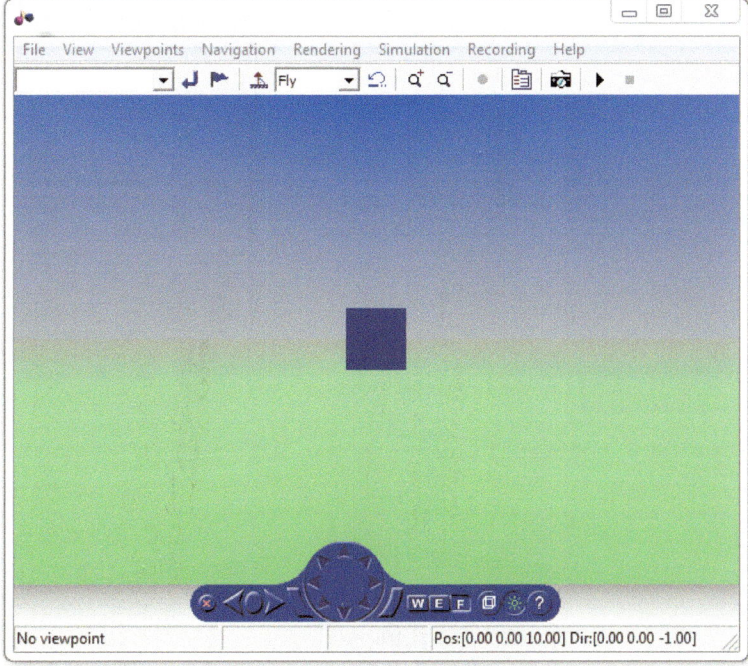

Fig. 8.8 Virtual scene

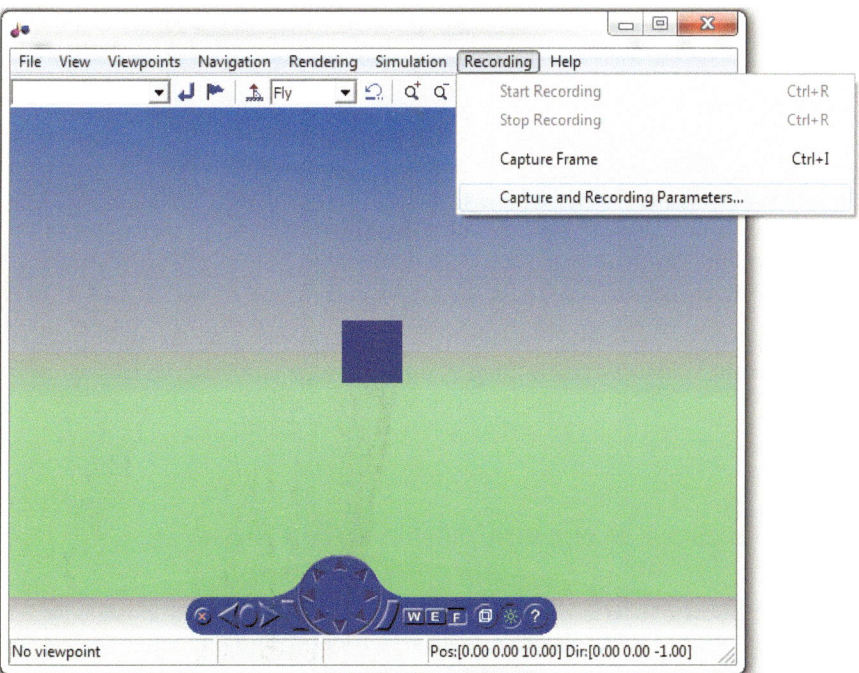

Fig. 8.9 Capture and recording parameters…

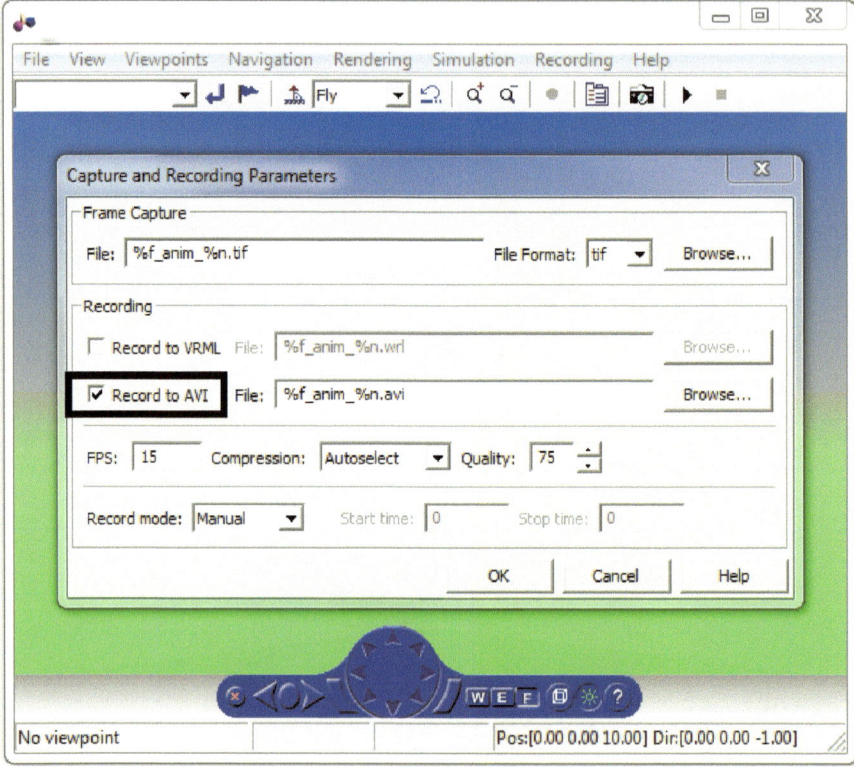

Fig. 8.10 The square to the left of record to AVI is shown in the rectangle

11. Click on **Start simulation** button to start the simulation (Fig. 8.11). In the working directory of MATLAB®, an animation video *Cube_Virtual_anim_1.avi* should be created. The user can play that movie to watch a recording of the animation.
12. Close the virtual scene and save the model.

For debugging purposes, the user can download the final Simulink® model (*Cube_animate_Simulink.mdl*) from Springer's web site (http://extras.springer. com/).

8.5   Application Problem: Collision of Two Boxes

Given two boxes of mass $m_{B1} = 1\,(\text{kg})$ and $m_{B2} = 2\,(\text{kg})$ and their dimensions are $2 \times 2 \times 2\,(\text{m}^3)$ and $4 \times 2 \times 2\,(\text{m}^3)$, respectively (Fig. 8.12). The initial velocity of box 1 (the smaller box) in the x-direction is $1\,(\text{m}/\text{s})$, while box 2 is initially at rest.

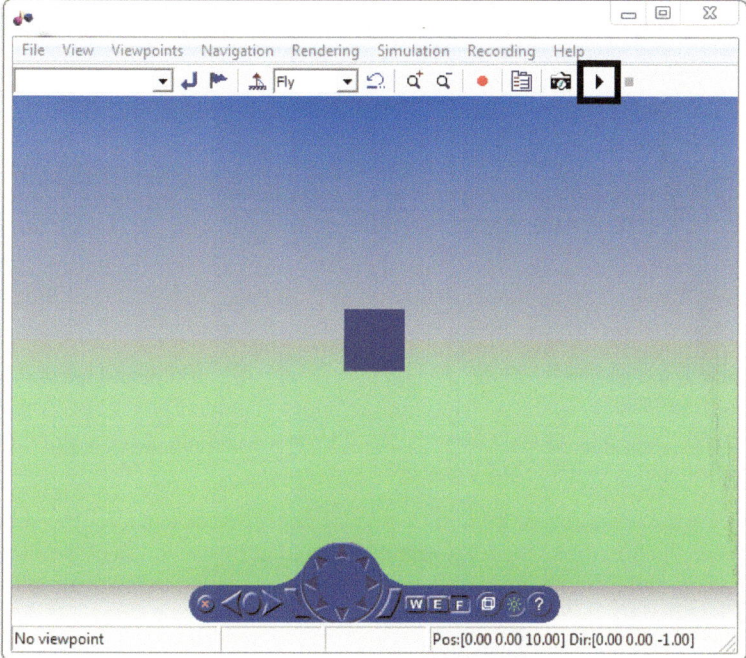

Fig. 8.11 Start simulation button

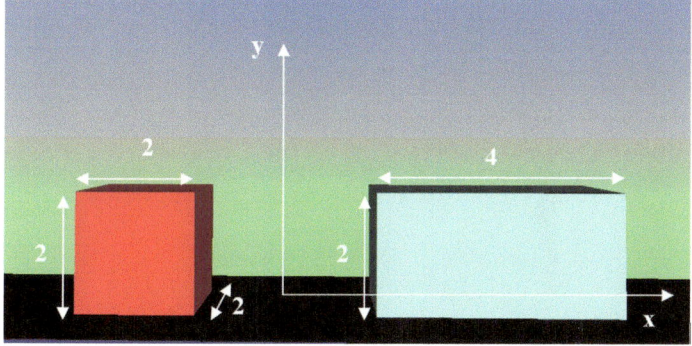

Fig. 8.12 Dimensions for the boxes

The initial position of box 1 is at $(-3,1,0)$, while box 2 is initially at $(3,1,0)$. Solve the following two parts (the solution is included in the electronic material of Chap. 8):

Part 1

Assuming no friction with the ground and using a coefficient of restitution of $e = 0.9$ (Hibbler 1986):

1. Reconstruct Fig. 8.12 in VRML assuming that the bottom faces of the boxes are located at $y=0$ and that the two boxes are aligned horizontally. Add an additional box component to represent the ground where the top face is at $y=0$.
2. Develop the equations of motion of box 1 and box 2 before and after the collision (Hibbler 1986).
3. Develop the Simulink® model to compute the numerical solution of the equations of motion and animate the VRML model. Assume that the total time of simulation is 10 s and that the step size is 0.005 s.
4. Plot the x-positions and x-velocities of boxes 1 and 2 as function of time. Plot the combined energy (potential and kinetic) of the system as function of time.

Part 2
Assume that the friction with the ground is opposing the motion and takes the form $-\mu \times (normal\ force)$. μ is the friction coefficient (Hibbler 1986) and is equal to 0.005 in this part, and *normal force* is equal to $m \times g$, where g is the gravitational acceleration and is equal to 9.81 m/s².

1. Develop the equations of motion of boxes 1 and 2 before and after the collision.
2. Develop the Simulink® model to compute the numerical solution of the equations of motion and animate the VRML model. Assume the total time of simulation is 10 s and that the step size is 0.005 s.
3. Plot the x-positions and x-velocities of boxes 1 and 2 as function of time. Plot the combined energy (potential and kinetic) of the system as function of time.
4. Compare the energy from Parts 1 and 2. What do you notice?

The solution of the problem can be downloaded from Springer's web site http://extras.springer.com/. The files can be found in folder /Chapter 8/Application Problem.

Reference

Book

Hibbler RC (1986) Engineering mechanics: dynamics, 4th edn. McMillan, New York

Chapter 9
Animation of Mass-Spring-Damper Oscillations Using Simulink®

9.1 Introduction

The purpose of this chapter is to educate Simulink® users on how to animate a physical system that has more than one component in a virtual reality environment. This chapter will walk the reader through a complete exercise on how to animate a physical problem that consists of a mass, a spring, and a damper that are governed by an equation of motion in VRML environment using a Simulink® model.

In Sect. 9.2, the mass-spring-damper problem will be presented where the mass is attached to a spring and subjected to an initial displacement. Newton's second law will be used to derive the equation of motion.

Section 9.3 will include a detailed description for constructing the Simulink® model that will numerically integrate for the equation of motion and will animate the virtual model, *virtual_scene.WRL*, for the mass-spring-damper that was created in Chap. 4.

The chapter concludes by an application problem of two masses connected by a spring. The problem will include developing the equations of motion of the two masses in addition to animating virtual scene of the physical system based on a set of governing differential equations.

The electronic version of all the Simulink® models, VRML files, in addition to the recorded movies for the problems can be downloaded from Springer's web site http://extras.springer.com/.

9.2 Mass-Spring-Damper Problem

Given a cube of mass $m = 1\,(\text{kg})$ and of dimensions $2 \times 2 \times 2\,(\text{m}^3)$ (Fig. 9.1). The cube is connected to a spring of stiffness $K = 1\,(\text{N/m})$ and a damping constant (friction) $C = 0.1\,(\text{Ns/m})$.

Matlab® and Simulink® are registered trademark of The Mathworks, Inc.

N. Khaled, *Virtual Reality and Animation for MATLAB® and Simulink® Users: Visualization of Dynamic Models and Control Simulations*, DOI 10.1007/978-1-4471-2330-9_9, © Springer-Verlag London Limited 2012

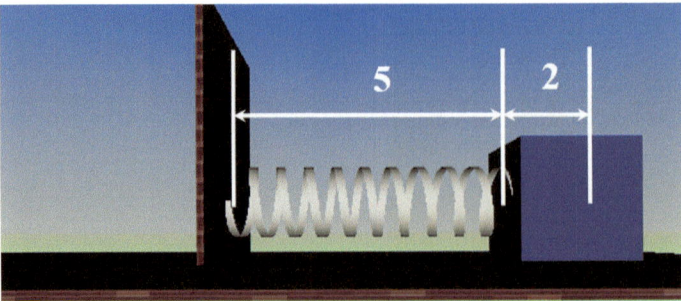

Fig. 9.1 Initial length of the spring in addition to the width of the mass

The motion of the system is constrained in the horizontal plane of the ground. The cube oscillates along the x-direction (horizontal direction of the screen). Using Newton's second law, the following second-order differential equation for the x-displacement of the mass can be developed:

$$m\ddot{x}(t) + C\dot{x}(t) + Kx(t) = 0 \Rightarrow \ddot{x}(t) = -\frac{K}{m}x(t) - \frac{C}{m}\dot{x}(t) \qquad (9.1)$$

The purpose of this exercise is to animate the solution of this equation for $\dot{x}(0) = 2$ (m/s) and $x(0) = 0$ (m). Note that the mass is oscillating about the equilibrium position shown in Fig. 9.1 (nonextended position of the spring).

9.3   Creating the Simulink® Model for Animating the Virtual Scene

In what follows, the virtual scene of the cube, *virtual_scene.WRL*, will be animated using Simulink®. The elements of the virtual scene that will be animated are the translation of the cube and the length of the spring. The translation property of the cube will be changed based on the solution of Eq. 9.1, and the length of the spring will be scaled according to the instantaneous position of the mass. The scale of the spring in the x-direction is given by:

$$scale\_x = (L_0 + x(t))/L_0 = (5 + x(t))/5 \qquad (9.2)$$

where L_0 is the original length of the spring.

Follow the following steps to create the Simulink® model that will animate the virtual scene (assuming that the user has the Virtual Reality Toolbox* installed in Simulink®):

1. Create a new Simulink® model.
2. Add a **VR Sink** from the **Virtual Reality Toolbox*** to the current model (Fig. 9.2). This sink will contain the virtual scene, *virtual_scene.WRL*, and will allow the user to manipulate the scene through Simulink® inputs.

---

*As of Matlab® 2009a, **Virtual Reality Toolbox** has been renamed **Simulink 3D Animation**.

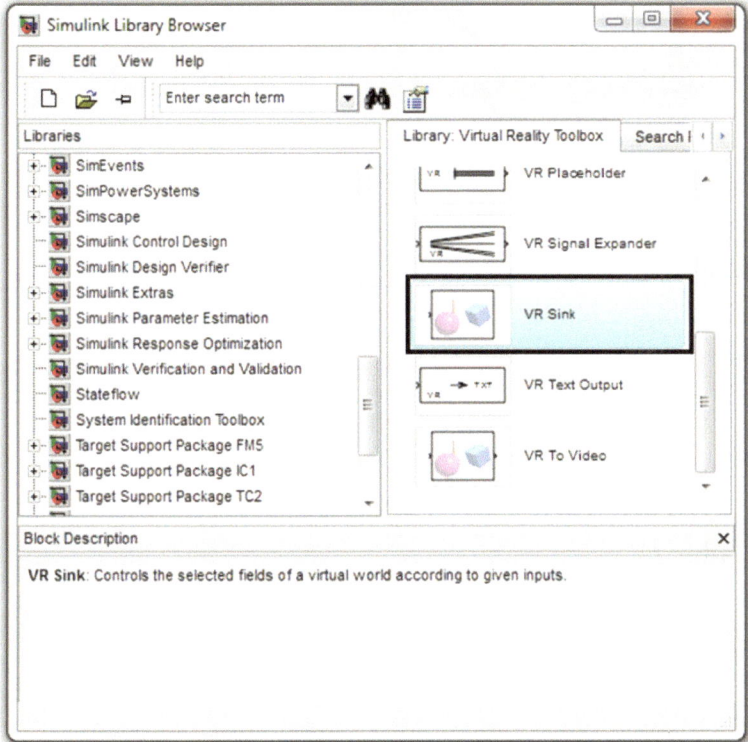

Fig. 9.2   VR Sink in Virtual Reality Toolbox*

3. Double-click the **VR Sink** in the model. A window for the parameters of the **VR Sink** will open (Fig. 9.3). Click on the **Browse** button (Fig. 9.3) to navigate and select the virtual scene *virtual_scene.WRL* which can be found in Chap. 9 material of the book. Download the file to the working directory of MATLAB®. Click **Open** once you select the file *virtual_scene.WRL*.

4. Change the **Sample time** of the **VR Sink** to 0.01(Fig. 9.4). Note that **Sample time** should match (or be a multiple of) the one set in the Simulink® model.

5. Click on the + sign to the left of **Mass (Transform)** (Fig. 9.5).

6. Tick the square to the left of **translation** property of **Mass** to select it (Fig. 9.6). This will enable the user to manipulate the **translation** property of the mass through Simulink® inputs.

7. Click on the + sign to the left of **Spring (Transform)** (Fig. 9.7).

8. Tick the square to the left of **scale** property of **Spring** to select it (Fig. 9.8). This will enable the user to manipulate the length of the spring through Simulink® inputs. Click **OK** to accept the changes and close the parameters of the **VR Sink**.

9. In the Simulink® model, go to **Simulation>Configuration Parameters** and change the solver type to **Fixed-step**. Change the **step size** to 0.01 to match the

---

*As of Matlab® 2009a, **Virtual Reality Toolbox** has been renamed **Simulink 3D Animation**.

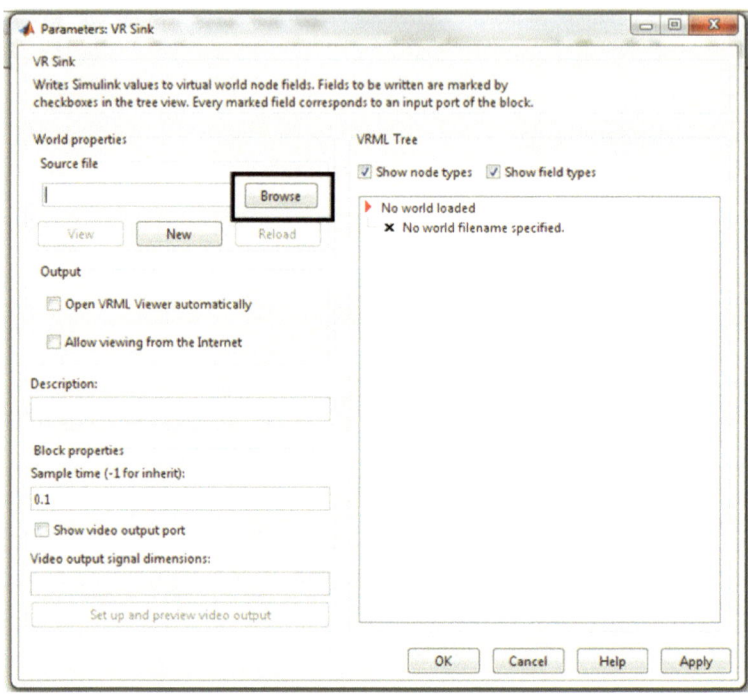

Fig. 9.3 Browse button for the virtual scene is shown in the rectangle

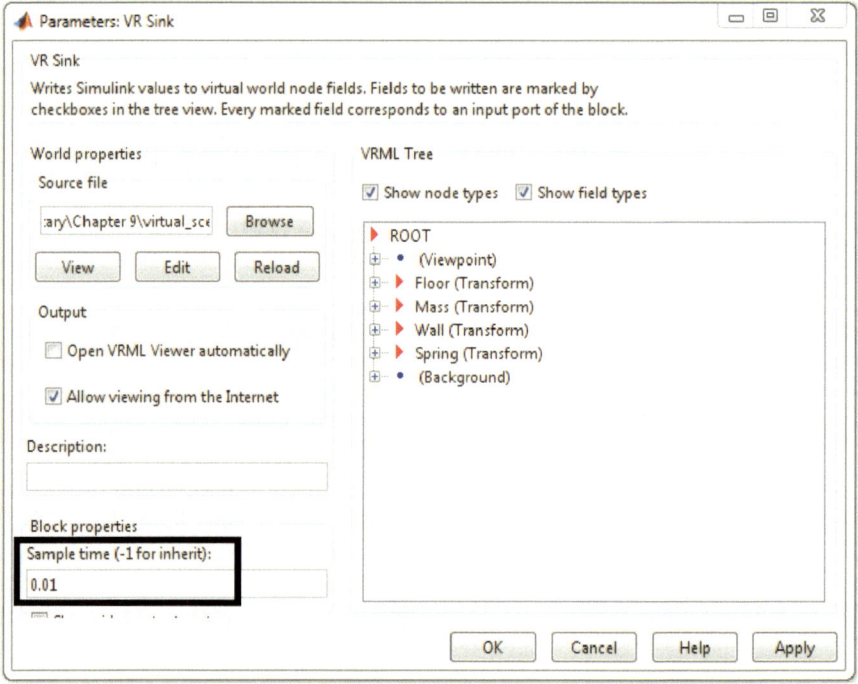

Fig. 9.4 Sample time

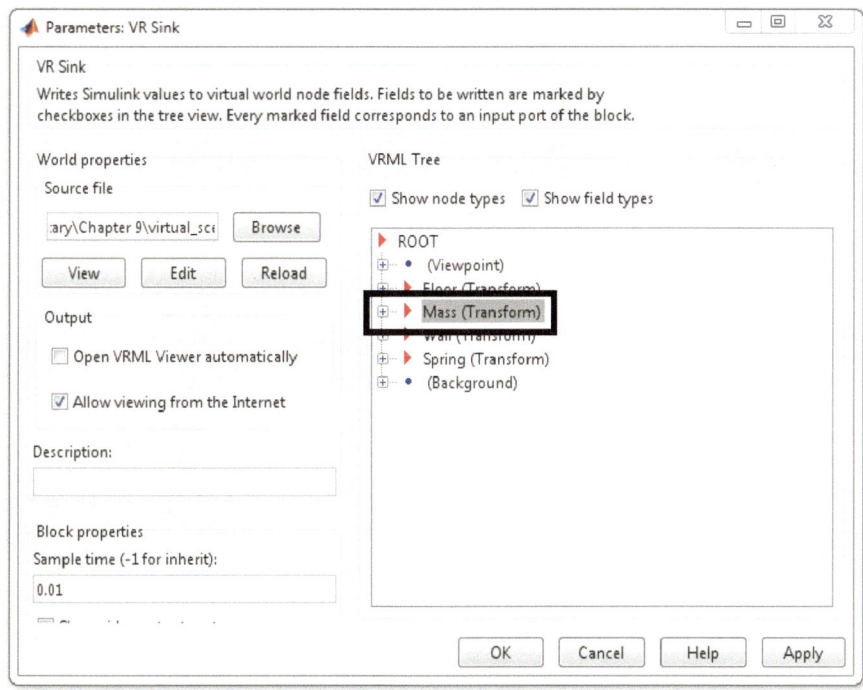

Fig. 9.5 Mass (Transform) is shown in the rectangle

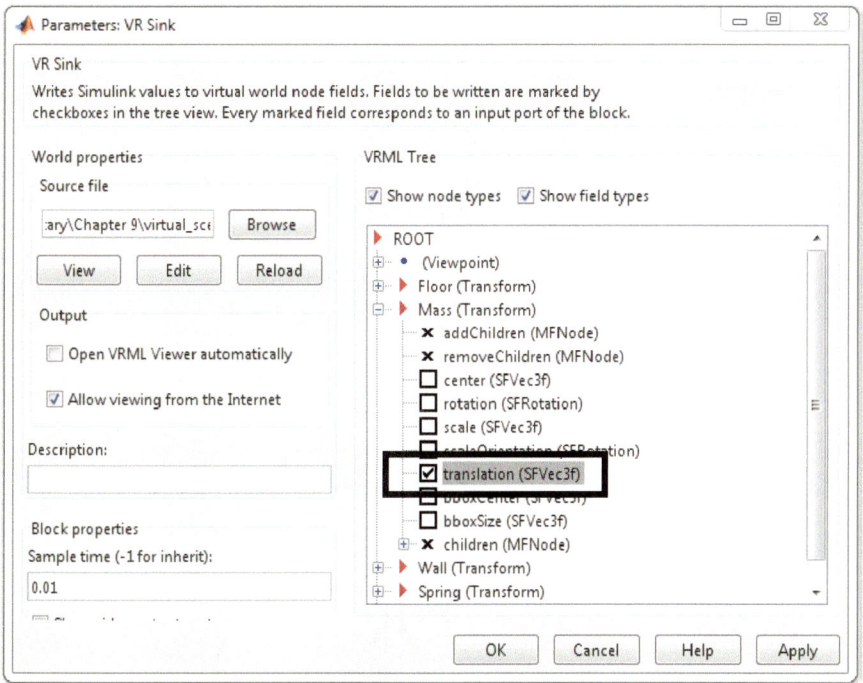

Fig. 9.6 The translation property is shown in the rectangle

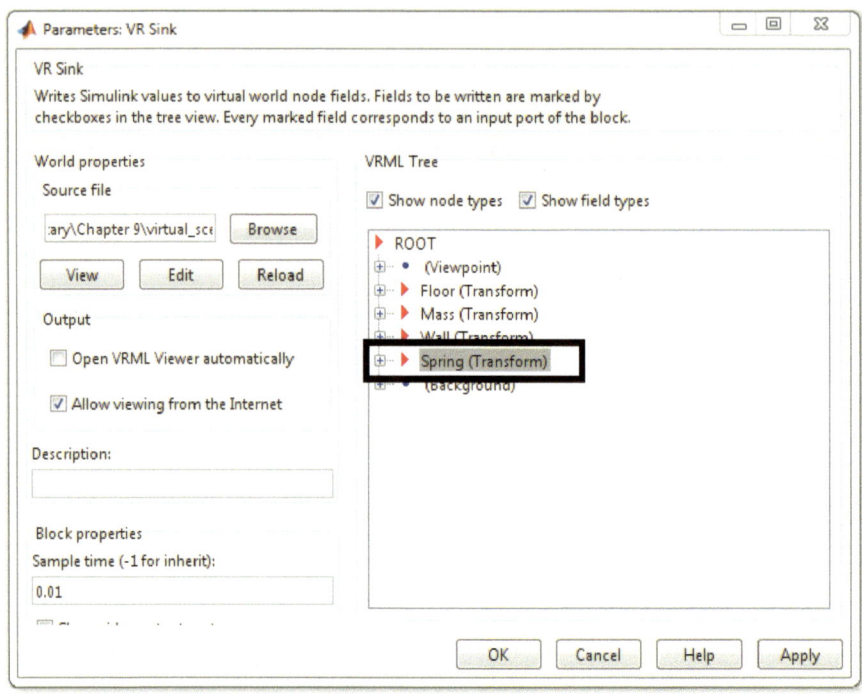

Fig. 9.7 Spring (Transform) is shown in the rectangle

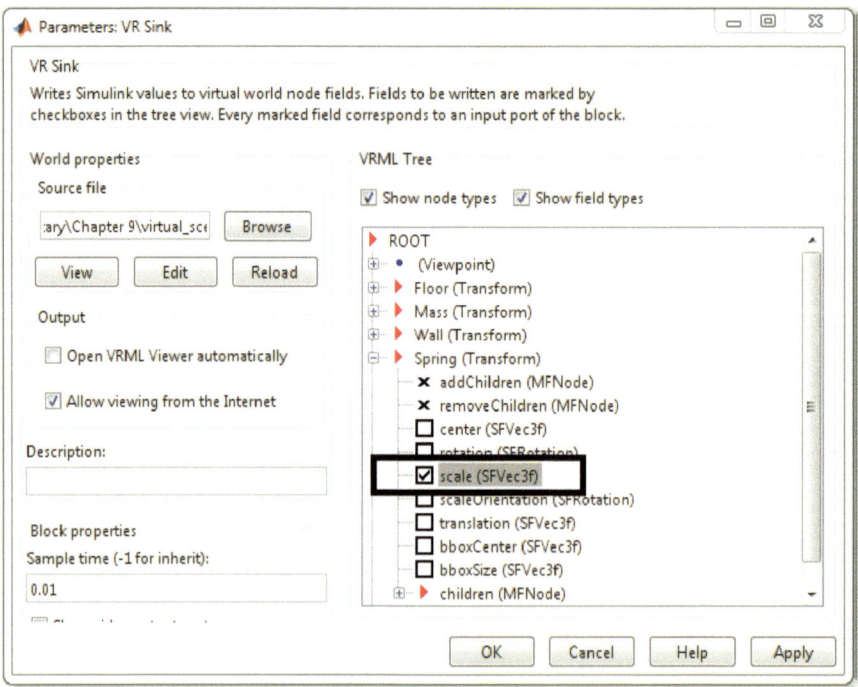

Fig. 9.8 The scale property is shown in the rectangle

Check File>Model Properties>Callbacks>InitFnc for the constants of the model

Based on Equation (9.1), compute the translation of the cube and the scale of the spring in the x direction

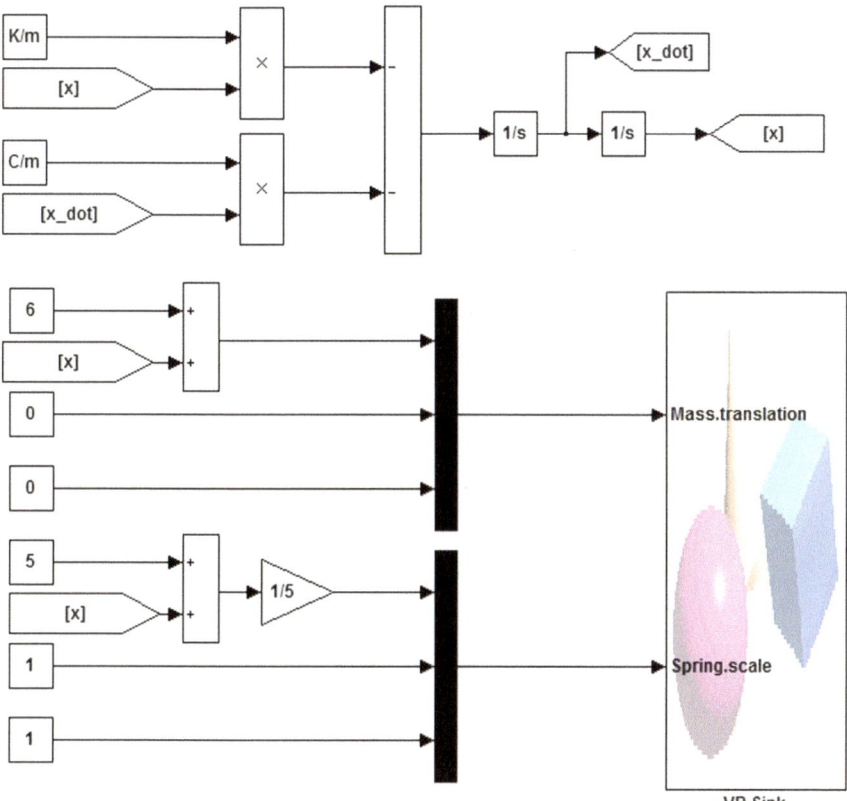

Fig. 9.9 Final model

Sample time of the **VR Sink**. In addition, change the **Stop time** to 50. Click **OK** when done.

10. Figure 9.9 shows the final model that is used to animate the virtual scene. The upper part of the model solves for (9.1) to translate the mass in the x-direction. Notice that a constant (6) is added to $x(t)$ since the oscillation of the mass is computed about its equilibrium position (refer to equation 9.2 for more details). Both **Mass.translation** and **spring.scale** properties are 3×1 signals.

11. Double-click the **VR Sink** to open the virtual scene (Fig. 9.10). Notice that the mass is defaulted at the origin of the coordinate system before the start of the simulation. This is the way the mass was defined in **VRML**. For more details, the reader is referred to Chap. 4.

12. Click on **Recording** and then choose **Capture and Recording Parameters…** to set up the recording options of the animated scene (Fig. 9.11).

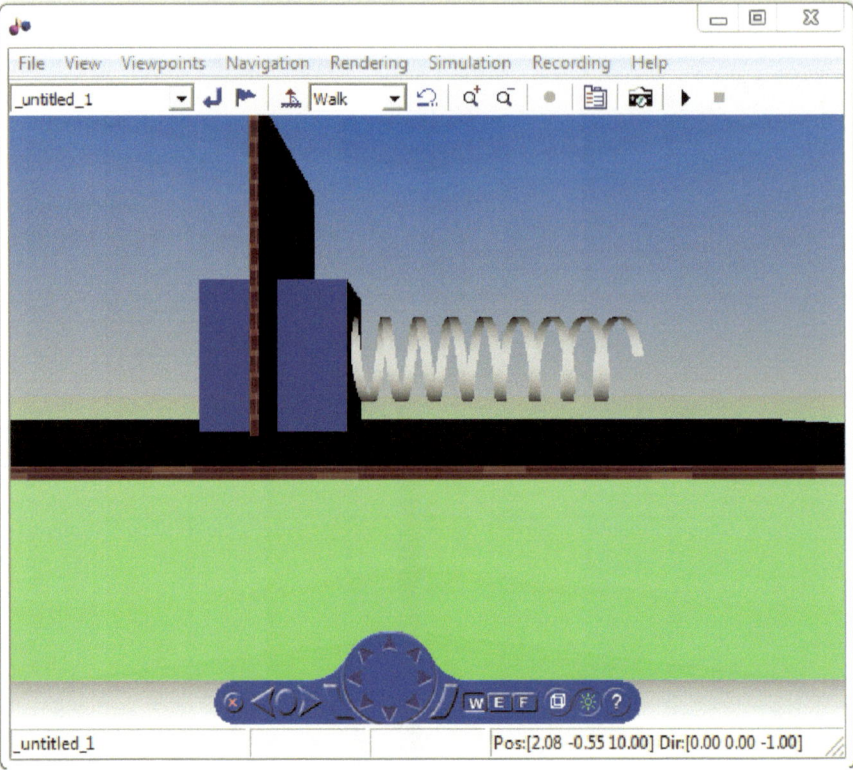

Fig. 9.10 Virtual scene

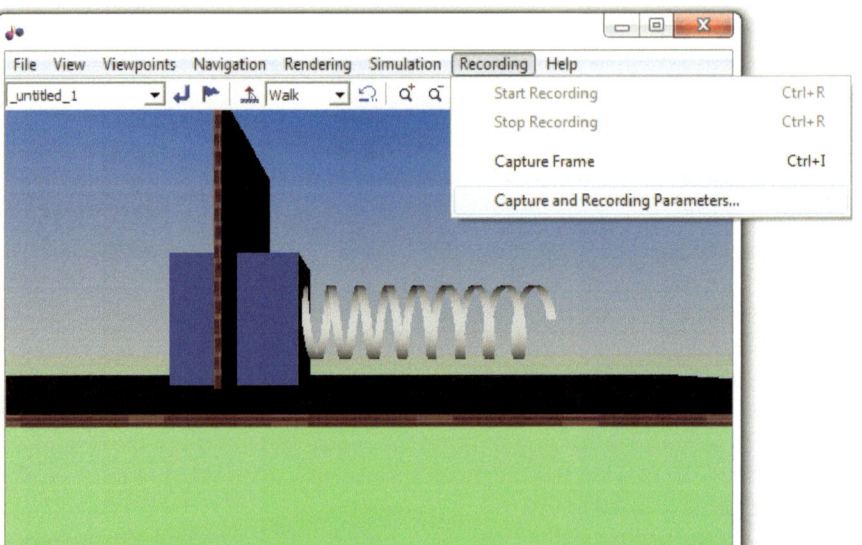

Fig. 9.11 Capture and Recording Parameters…

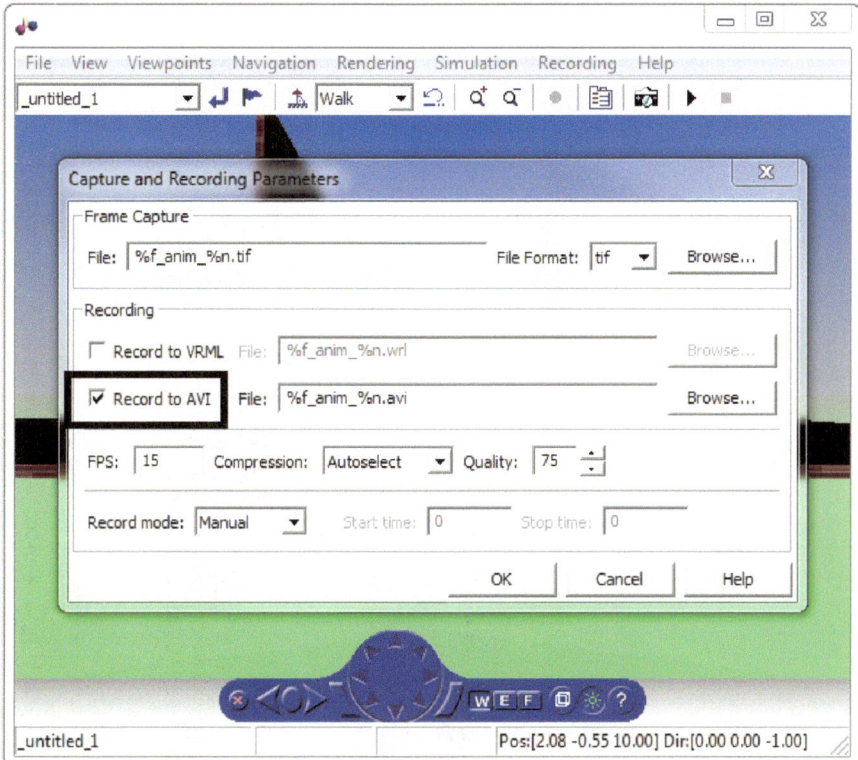

Fig. 9.12 Record to AVI is shown in the rectangle

13. To record the animation, tick the square to the left of **Record to AVI** (Fig. 9.12). The first time the user runs and records the animation, the default name of the animation will be saved to *virtual_scene_anim_1.avi* (*virtual_scene.WRL* is the name of the virtual scene). Index 1 will automatically be incremented each time the user reruns the simulation; thus, recordings of multiple simulations will be saved rather than overwritten.

14. To set up the start and stop time of the recording of the animation, change the **Record mode** to **Scheduled** and set the **Stop time** to 50 which is the stop time of the simulation (Fig. 9.13). Click **OK** when done.

15. Click on **Start simulation** button to start the simulation. In the working directory of MATLAB®, an animation video *virtual_scene_anim_1.avi* should be created. The user can play the movie to watch a recording of the animation. If the movie is running too slow, the user can go back to steps 4 and 9 and increase the step size.

16. Close the virtual scene and save the model.

For debugging purposes, the user can download the final Simulink® model (*MSD. mdl*) from Springer's web site (http://extras.springer.com/).

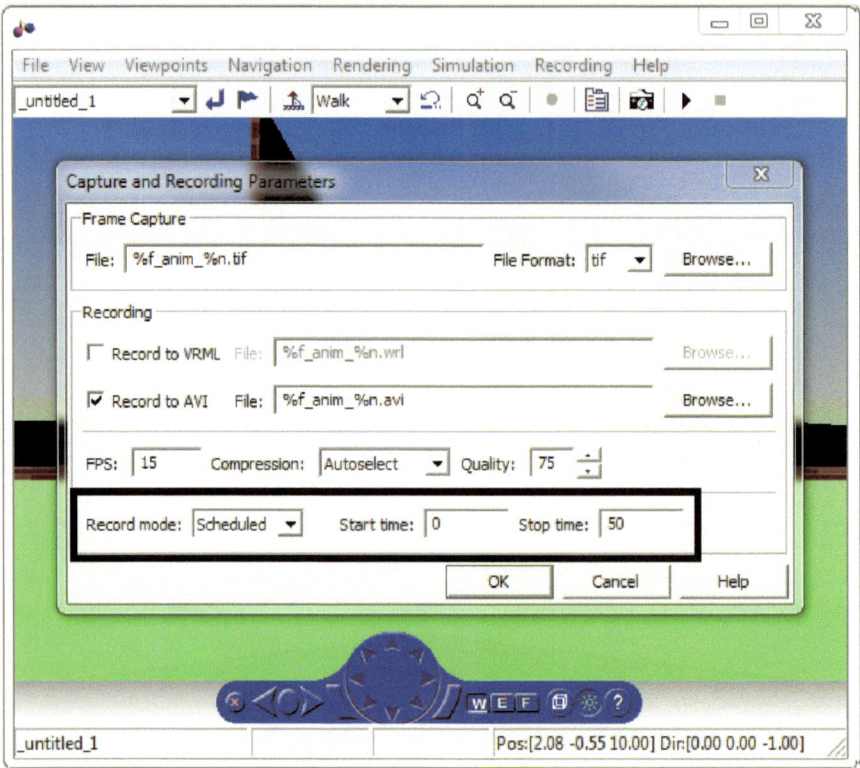

Fig. 9.13 Record mode and Stop time are shown in the rectangle

9.4   Changing the Viewpoint in Simulink®

When running the simulation, the user might want to view the animation from a different **Viewpoint** (point where the viewer is looking at the scene). The user can change the **Viewpoint** either in V-Realm Builder or in Simulink®. To change the **Viewpoint** in Simulink®, double-click the **VR Sink** to open the virtual scene. Notice the **W**, **E**, and **F** buttons to the right of the **Navigation Wheel** (Fig. 9.14). If the **W** button is selected (usually it is selected by default), the user can make the view closer (further) by clicking the up (down) arrow of the **Navigation Wheel**. In addition, the view can be rotated to the right (left) direction by clicking the right (left) arrow of the **Navigation Wheel**. For more detailed information, the interested reader can click **Help** on the top right corner of the virtual scene to explore the **E** and **F** buttons.

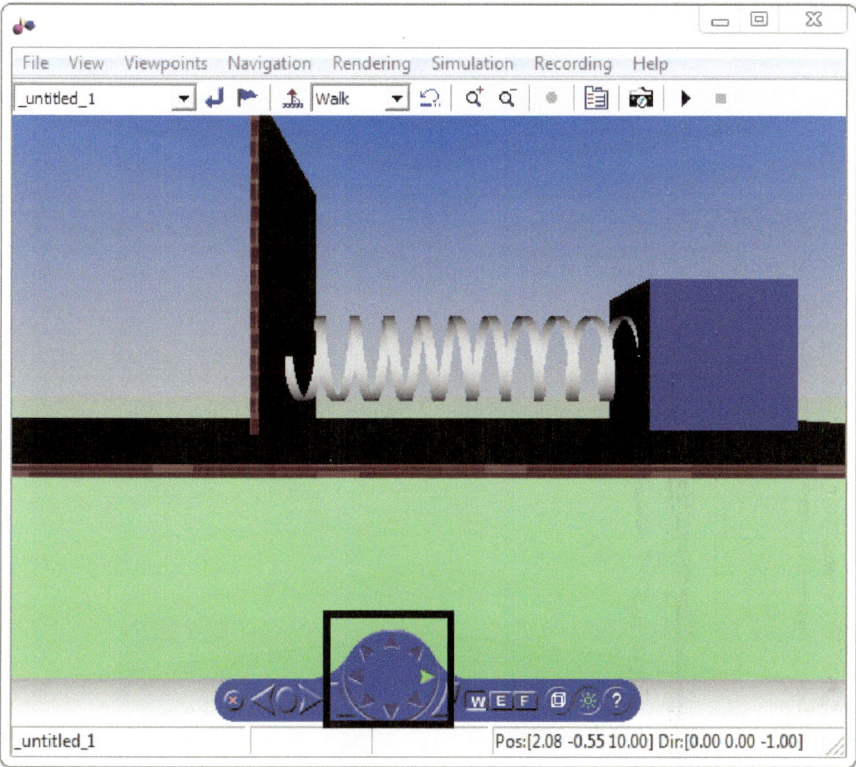

Fig. 9.14 Navigation Wheel is shown in the rectangle

9.5 Application Problem: Collision of Two Boxes

Given two masses $m_{B1} = 1\,(\text{kg})$ and $m_{B2} = 1\,(\text{kg})$ of the same dimensions $1 \times 2 \times 2\,(\text{m}^3)$ (Fig. 9.15). The two masses are connected by a spring of stiffness $K = 1\,(\text{N/m})$ and a damping constant (friction) $C = 0.15\,(\text{Ns/m})$. The radius of the unextended spring is $0.4\,(\text{m})$, and the length is $5\,(\text{m})$ (Fig. 9.15). The spring is initially compressed, and the centers of m_{B1} and m_{B2} are initially situated at $x_{B1}(0) = -1.7\,(\text{m})$ and $x_{B2}(0) = 1.7\,(\text{m})$, respectively (Fig. 9.16). Both masses have zero initial velocities. The virtual scene, ***mass_1_mass_2.WRL***, can be downloaded from Chap. 9 material. Assuming no friction with the ground, solve the following:

1. Develop the equations of motion of the two masses (Hibbler 1986).
2. Develop the Simulink® model to compute the numerical solution of the equations of motion, animate the VRML model, and record the simulation into a video. The objects that should be animated in the scene are the two masses and the spring.

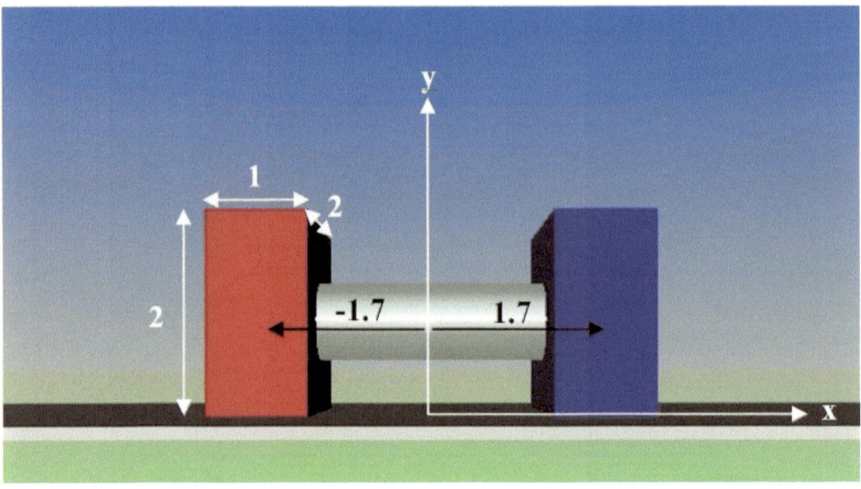

Fig. 9.15 Dimensions of the two masses and their initial positions

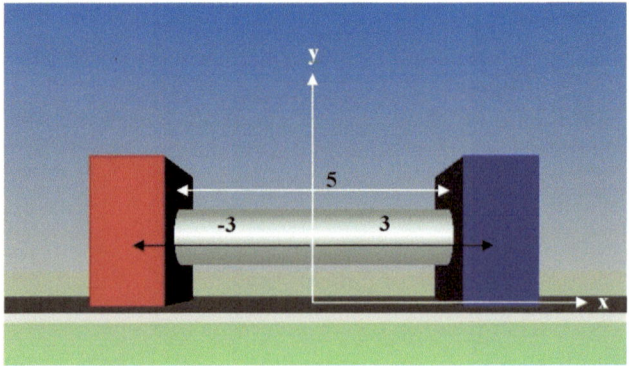

Fig. 9.16 Unextended position of the spring with the two masses

The spring should be scaled in the x-direction based on the instantaneous positions of the two masses on the condition that its initial volume does not change. Assume that the total time of simulation is 50 s and that the step size is 0.01 s. Hint: From VRML model, the x-scale of the spring represents the radius scale, and the y-scale represents the height scale. Keep the z-scale equal to 1.

The solution of the problem can be downloaded from Springer's web site http:// extras.springer.com/. The files can be found in folder /Chapter 9/Application Problem.

Reference

Book

Hibbler RC (1986) Engineering mechanics: dynamics, 4th edn. McMillan, New York

Chapter 10
Animation of Crank-Slider Mechanism of a Piston Using Simulink®

10.1 Introduction

The purpose of this chapter is to give Simulink® users a better understanding on how to animate a physical system that has more than one component in a virtual reality environment. This chapter will also help the reader to implement a simple PID controller to control a ball on a plate and visualize the performance of the controller.

In Sect. 10.2, a crank-slider mechanism problem will be presented.

Section 10.3 will describe in details how to construct the Simulink® model that will animate the virtual scene.

Section 10.4 will guide and teach the user how to change the **Viewpoint** of the virtual scene in Simulink® before running the simulation.

The chapter concludes by a control problem for a ball on a platform. The position of the ball is controlled on the platform by means of two independent actuators. The problem will include developing the equations of motion of the ball, developing the two PID controllers for the platform, and animating the ball and the platform.

The electronic version of the Simulink® models, VRML models, in addition to the recorded movies for the animated crank-slider mechanism and the controlled motion of the ball on the plate can be downloaded from Springer's web site http://extras.springer.com/.

10.2 Crank-Slider Mechanism

Figure 10.1 shows a sketch for the crank-slider mechanism. Point O is the center of the crank shaft. ON is the crank shaft. Its length is $r = 10$ (cm). NP is the connecting rod. Its length is $l = 20$ (cm). Since P is constrained to move in the vertical direction

---

Matlab® is a registered trademark of The Mathworks, Inc.

N. Khaled, *Virtual Reality and Animation for MATLAB® and Simulink® Users:*
Visualization of Dynamic Models and Control Simulations,
DOI 10.1007/978-1-4471-2330-9_10, © Springer-Verlag London Limited 2012

Fig. 10.1 A sketch for the
crank-slider mechanism

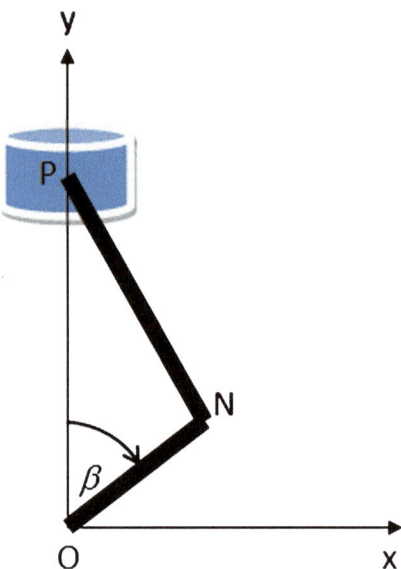

and the rest of the components are coplanar, this system has one degree of freedom which will be considered to be the crankshaft angle β.

The coordinates for point O, which is the pivot point for the crankshaft, are $(0,0,0)$. The coordinates of point N are $(r\sin(\beta), r\cos(\beta), 0)$. The coordinates of point P are $(0, r\cos(\beta) + \sqrt{l^2 - r^2 \sin^2(\beta)}, 0)$.

10.3 Creating the Simulink® Model for Animating the Virtual Scene

In what follows, the virtual scene, ***virtual_scene.wrl***, will be animated by assuming that the crank angle is an input to the system. Figure 10.2 shows the centers (midpoints) M_1 and M_2 of the crank and the connecting rod, respectively. In addition, Fig. 10.2 shows β and γ, the angles of the crank and the connecting rod with the vertical, respectively. The parameters that will be changed in the virtual world are:

1. The **translation** property of **Piston**
2. The **rotation** property of the **Rod** (around its center M_2)
3. The **translation** property of **Rod** (translation of its center M_2)
4. The **rotation** property of **Crank** (around its center M_1)
5. The **translation** property of **Crank** (translation of its center M_1)

The crank will be assumed to be rotating in a constant speed, $\omega = 0.1$ rad/s (intentionally chosen small to enable the user to track the motion). To compute the crank angle, integrate ω with respect to time $\beta = \int_0^t \omega d\tau$. As for the translation of the

Fig. 10.2 M_1 and M_2 are the centers (midpoints) of the crank and connecting rod, respectively

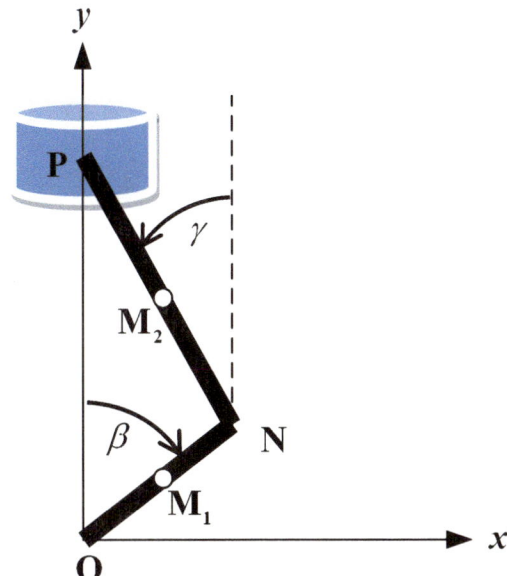

crank ON, it is represented by the midpoint of the crank, M_1 (Fig. 10.2), and its coordinates are given by [coordinates of N]/2. The rotational angle of the crank is set to $-\beta$ for clockwise rotation. As for the translation of the rod NP, it is represented by the midpoint of the rod, M_2 (Fig. 10.2), and its coordinates are given by [(coordinates of N)+(coordinates of P)]/2. The rotational angle of the rod is given by atan2[(P(1) − N(1)),(P(2) − N(2))]. The piston is only undergoing translational motion in the vertical direction, and its translation is represented by the coordinates of P.

Follow the steps given below to create the Simulink® model that will animate the virtual scene (assuming that the user has the **Virtual Reality Toolbox*** installed in Simulink®):

1. Create a new Simulink® model.
2. Add a **VR Sink** from the **Virtual Reality Toolbox*** to the current model (Fig. 10.3). This sink will contain the virtual scene *virtual_scene.wrl* and will allow the user to manipulate the scene through Simulink® inputs.
3. Double-click the **VR Sink** in the model. A window for the parameters of the **VR Sink** will open (Fig. 10.4). Click on the **Browse** button (Fig. 10.4) to navigate and select the virtual scene *virtual_scene.wrl* which should be in Chap. 10 material of the book. Download the file to the working directory of MATLAB®. Click **Open** once you select the file *virtual_scene.wrl*.
4. Keep the **Sample time** of the **VR Sink** 0.1(Fig. 10.5). Note that **Sample time** should be equal to (or a multiple of) the one set in the Simulink® model.

---

* As of Matlab® 2009a, **Virtual Reality Toolbox** has been renamed **Simulink 3D Animation**.

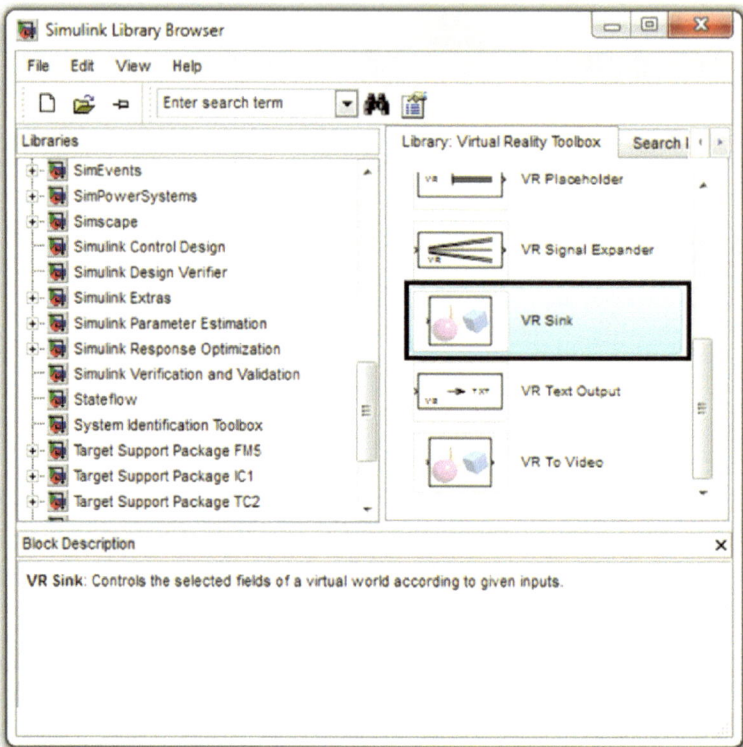

Fig. 10.3 VR Sink in Virtual Reality Toolbox*

5. Click on the + sign to the left of **piston (Transform)** (Fig. 10.6).
6. Tick the squares to the left of **translation** property of **piston** to select it (Fig. 10.7). This will enable the user to manipulate the translation of the **piston** through Simulink® inputs.
7. Click on the + sign to the left of **rod (Transform)** (Fig. 10.8).
8. Tick the squares to the left of **translation** and **rotation** properties of **rod** to select them (Fig. 10.9). This will enable the user to manipulate the translation and rotation of the **rod** through Simulink® inputs.
9. Click on the + sign to the left of **crank (Transform)** (Fig. 10.10).
10. Tick the squares to the left of **translation** and **rotation** properties of **crank** to select them (Fig. 10.11). This will enable the user to manipulate the translation and rotation of the **crank** through Simulink® inputs. Click **OK** when done.
11. In the Simulink® model, go to **Simulation>Configuration Parameters** and change the solver type to **Fixed-step** (Fig. 10.12). Note that the **step size** should be set to 0.1 to match the **Sample time** of the **VR Sink**. In addition, change the **Stop time** to 100*pi. Click **OK** when done.

---

*As of Matlab® 2009a, **Virtual Reality Toolbox** has been renamed **Simulink 3D Animation**.

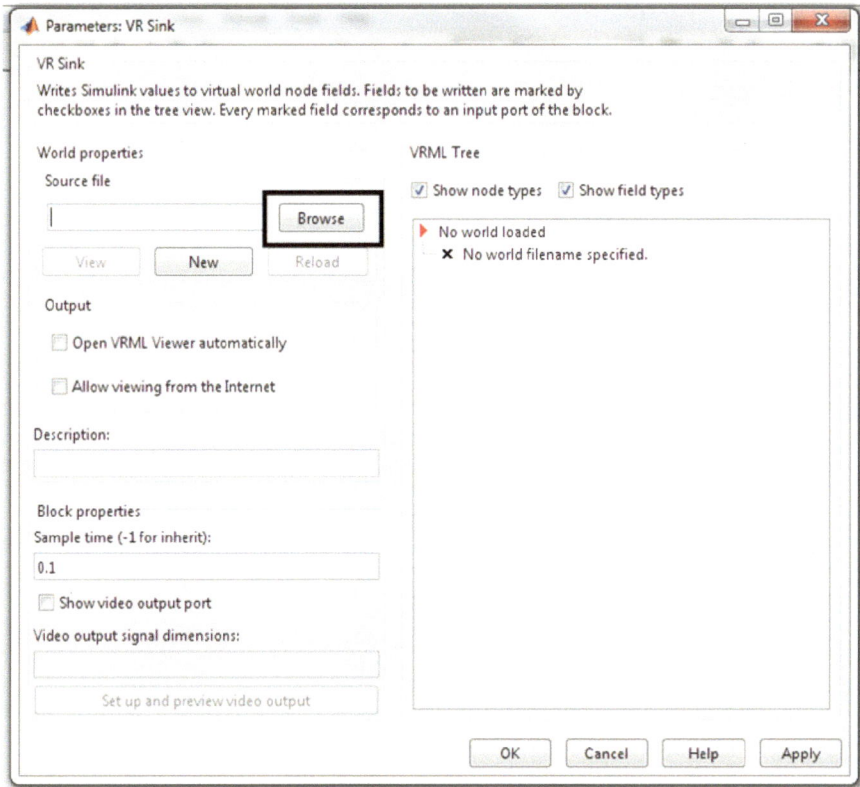

Fig. 10.4 Browse button for the virtual scene is shown in the rectangle

12. Figures 10.13, 10.14 and 10.15 show the final model that is used to animate the virtual scene. It is worth mentioning that the translation input to the **VR Sink** is 3×1 and the rotation input is 4×1 vectors.
13. Double-click the **VR Sink** to open the virtual scene (Fig. 10.16). Notice that the piston, the rod, and the crank are defaulted at the origin of the coordinate system before the start of the simulation. As for the cylinder, it was translated in VRML. For more details, the reader is referred to Chap. 5.
14. Click on **Start simulation** (Fig. 10.16) button to start the simulation.

To compare and debug the final model, the user can check the model **Crank_slider_mechanism.mdl** in Chap. 10 that can be downloaded from Springer's web site http://extras.springer.com/.

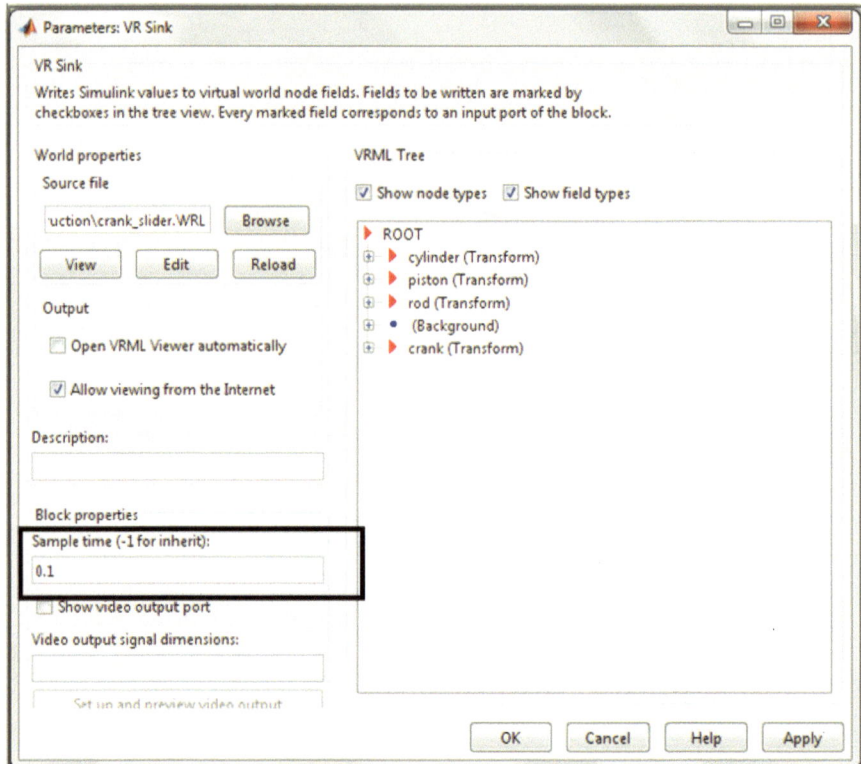

Fig. 10.5 Sample time for VR Sink

10.4   Changing the Viewpoint in Simulink®

There are two ways to change the **Viewpoint**. The user can either create various **Viewpoints** in VRML when constructing the scene and select the desired **Viewpoint** before running the simulation in Simulink® or he/she can open the virtual scene before running the simulation and manipulate the view by using the mouse. In this section, the latter will be used to change the **Viewpoint**. Open the model created in the previous section and double-click the **VR Sink**. Click the Examine button (Fig. 10.17). Click and hold the left button of the mouse. The cursor will change to a cross (Fig. 10.18). Slowly drag the figure down to rotate the view (Fig. 10.19). Now you can run the simulation with the new view.

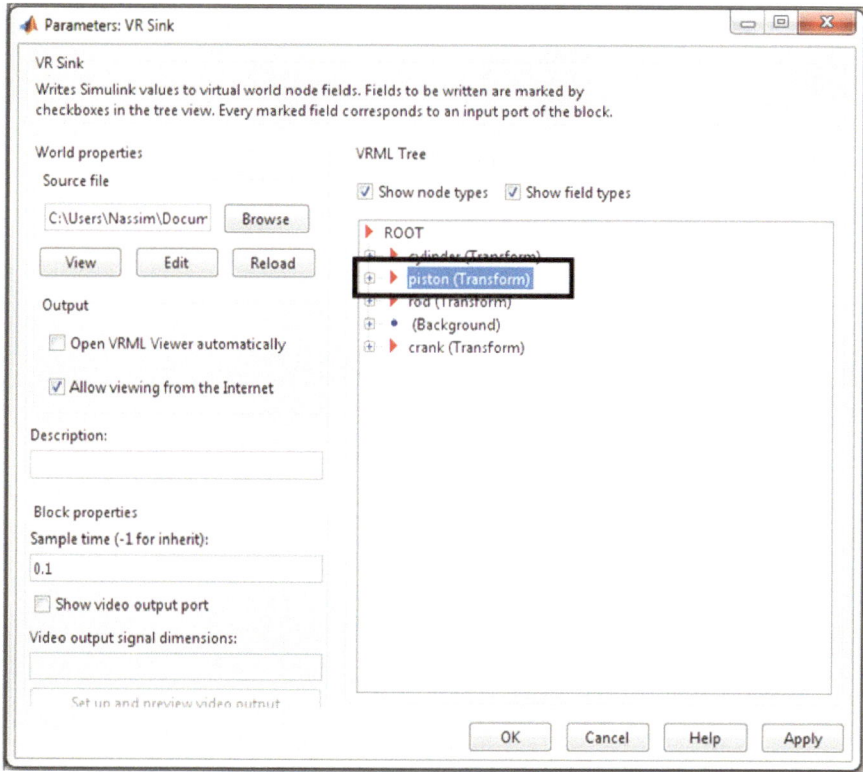

Fig. 10.6   The piston (Transform) is shown in the rectangle

10.5   Application Problem: Control of a Ball on a Plate

Given a ball of mass $m = 0.008$ (kg) and radius $r_0 = 0.5$ (m) (Fig. 10.20). The plate is of dimensions $7 \times 0.2 \times 7$ (m^3). The plate is pivoted around the origin $O(0,0,0)$ and has two independent actuators that can rotate it around x and z global axes by angles β and α, respectively ($-0.5 \text{ rad} \leq \beta \leq 0.5 \text{ rad}$ and $-0.5 \text{ rad} \leq \alpha \leq 0.5 \text{ rad}$). Assuming no friction and no slipping between the plate and the ball, solve the following:

1. Develop two equations of motion of the 2 degrees of freedom ball (Hibbler 1986). Where I_0 is the moment of inertia of the ball, prove that the equations of motions are:

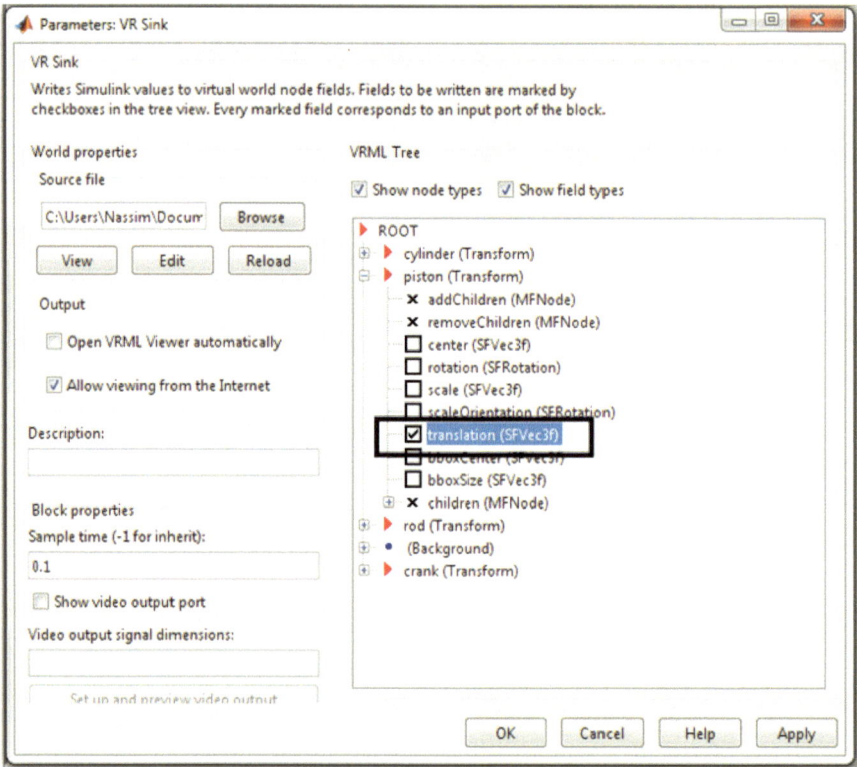

Fig. 10.7 The translation property of Piston

$$\left(m+\frac{I_0}{r_0^2}\right)\ddot{x} - m\left(\dot{\alpha}\dot{\beta}z + \dot{\alpha}^2 x\right) + mg\sin(\alpha) = 0 \qquad (10.1)$$

$$\left(m+\frac{I_0}{r_0^2}\right)\ddot{z} - m\left(\dot{\alpha}\dot{\beta}z + \dot{\beta}^2 x\right) + mg\sin(\beta) = 0 \qquad (10.2)$$

2. Neglecting the actuator dynamics (i.e., assuming a commanded angle α or β can be instantly delivered by the actuators) and assuming that no coupling between the two actuators, develop the two independent PID controllers (Ogata 2010) for the angles α and β of the plate to bring the ball to origin within a settling time of 50 and 70 s for x and z, respectively (hint: Define the error for α and β controllers to be $(x-0)$ and $(z-0)$, respectively). Develop the Simulink®

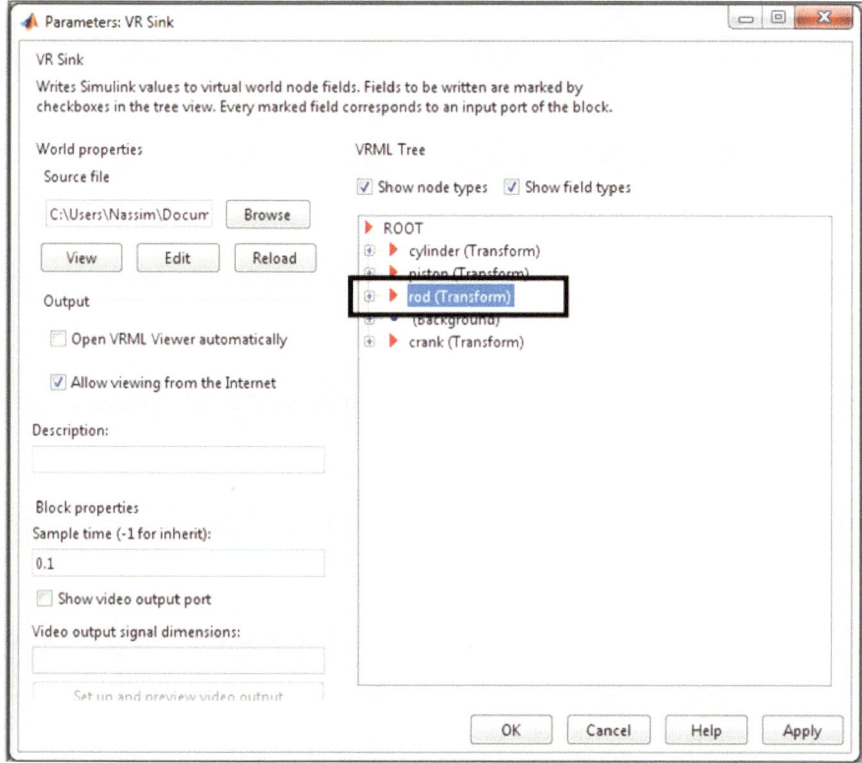

Fig. 10.8 The rod (Transform) is shown in the rectangle

model to simulate the two controllers along with numerical integration of Eqs. 10.1 and 10.2. Set the total time of simulation to 80 s and the step size to 0.05 s. The initial conditions for system are:

$$x(0) = 1.3 \, (\text{m})$$

$$z(0) = -2.4 \, (\text{m})$$

$$\dot{x}(0) = 0.1 \, (\text{m/s})$$

$$\dot{z}(0) = 0.3 \, (\text{m/s})$$

$$\alpha(0) = 0 \, \text{rad}$$

$$\beta(0) = 0 \, \text{rad}$$

3. Add a **VR Sink** to the model to animate virtual scene *Ball_platform.wrl* (Fig. 10.21) for the ball and platform system. The virtual scene can be downloaded from Springer's web site http://extras.springer.com/. The virtual scene can be

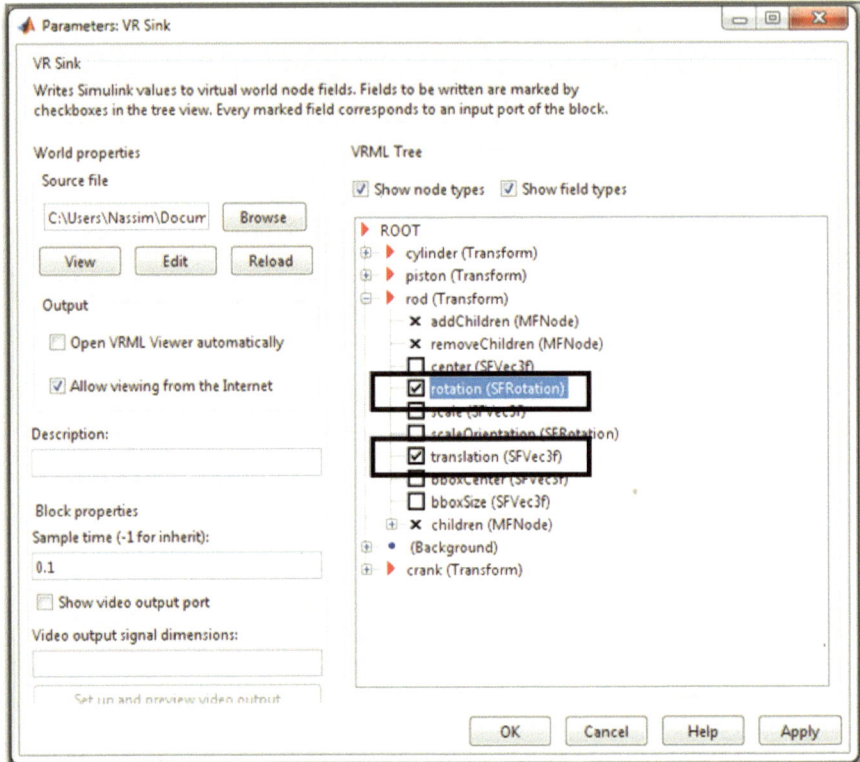

Fig. 10.9   The rotation and translation properties of rod

found in folder /Chapter 10/Application Problem. Follow the following hints to animate the scene:

- The reader has to construct the rotation matrix of the platform by using the two rotation angles α and β.
- The reader has to watch out for the sign convention for the rotation angles (α and β) when constructing the rotation matrix. From Fig. 10.20, a positive rotation β of the plate around x-axis would ultimately yield a positive z-displacement of the ball. Similarly, a positive rotation α around the z-axis would ultimately yield a positive x-displacement of the ball. β is oriented anticlockwise for its positive direction, while α is oriented clockwise for its positive direction.

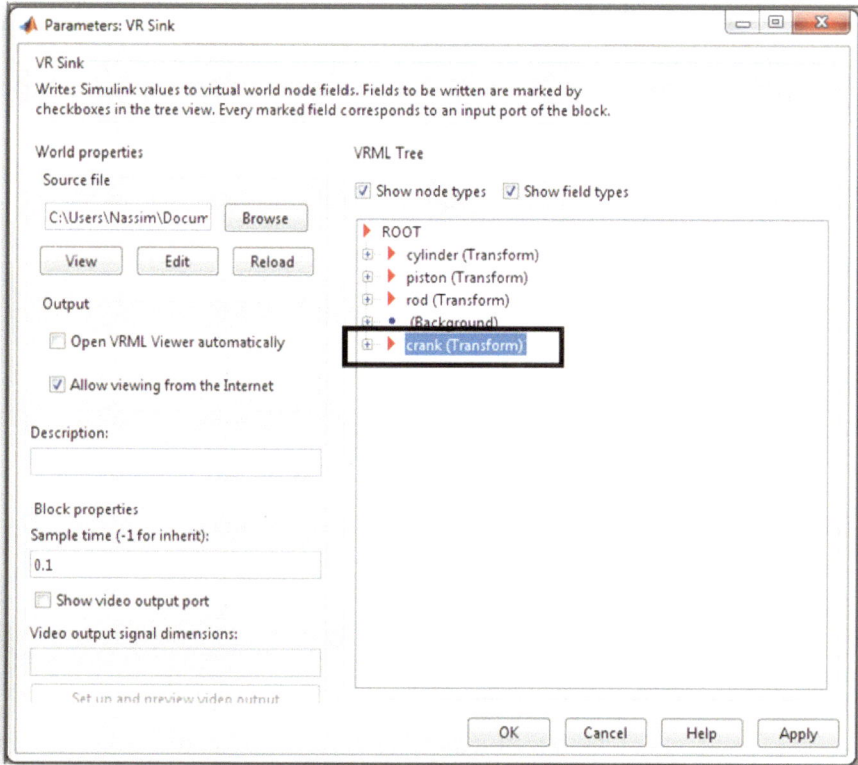

Fig. 10.10 The crank (Transform) is shown in the rectangle

- Once the rotation matrix has been constructed, the reader is advised to use the **Rotation Matrix to VRML Rotation** block from the **Virtual Reality Toolbox** block to change the rotation matrix (3×3) to a rotation vector (4×1). The rotation vector can be used as an input to the VRML model to rotate the plate.
- The reader has to compute the global y-position of the ball. To do so, the reader can use the normal vector of the plate in addition to the fact that the center of the ball is always at a perpendicular distance equal to the radius of the ball. Once y is computed, the position of the ball in space is changed in VRML by changing the translation property of the ball to (x, y, z).

The solution of the problem can be downloaded from Springer's web site http://extras.springer.com/. The files can be found in folder /Chapter 10/Application Problem.

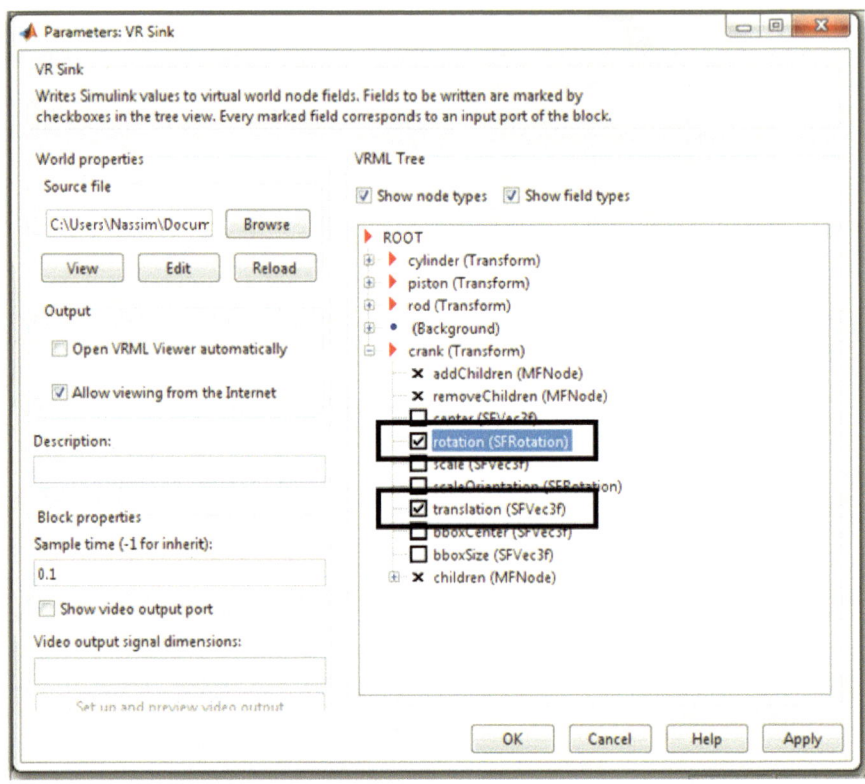

Fig. 10.11 The rotation and translation properties of crank

Fig. 10.12 Solver options

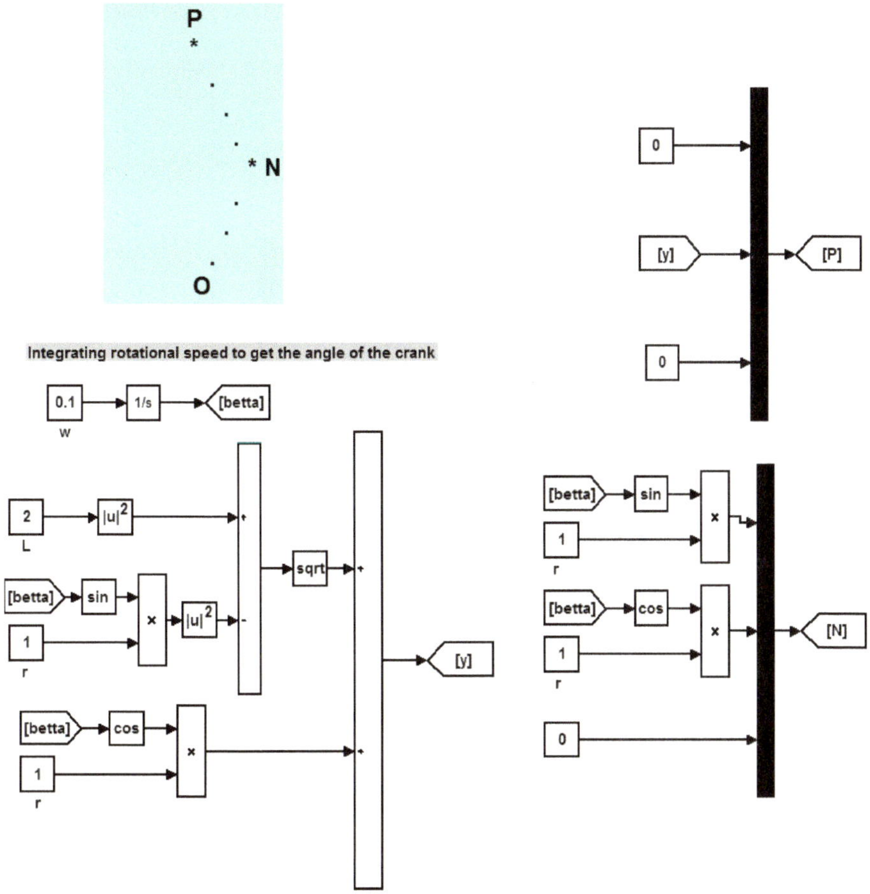

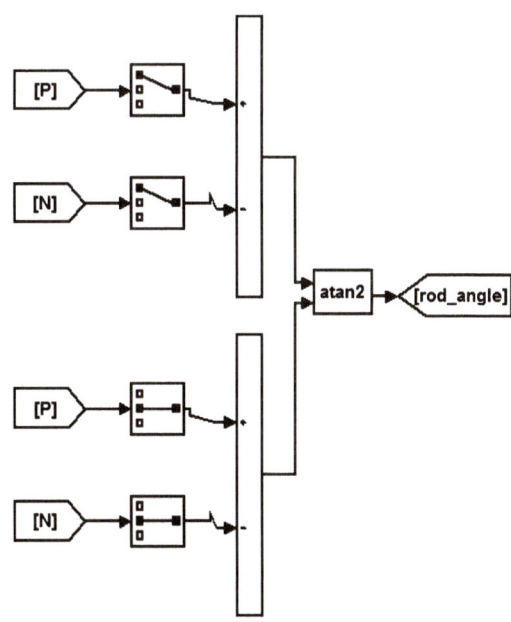

Fig. 10.13 Coordinates of N and P

Fig. 10.14 The rod angle

Fig. 10.15 VR Sink with all
the inputs

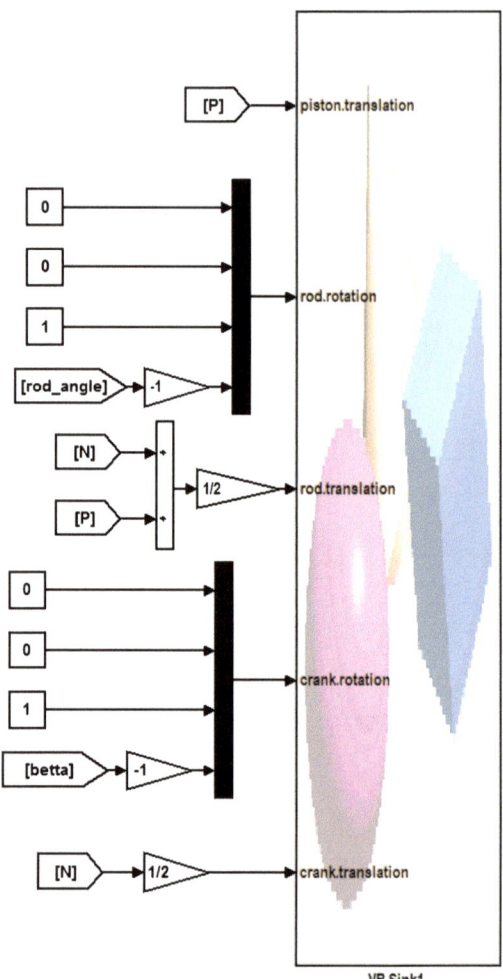

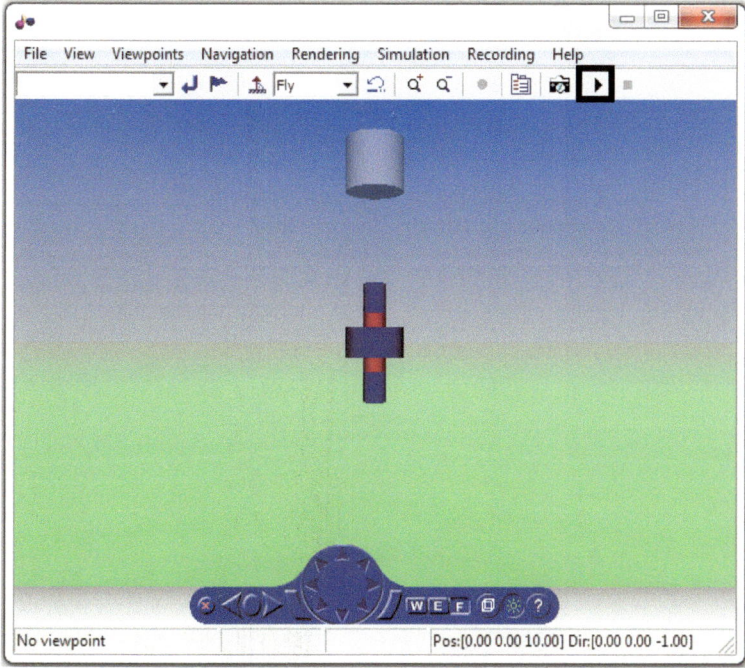

Fig. 10.16 Virtual scene and the Start simulation button in the rectangle

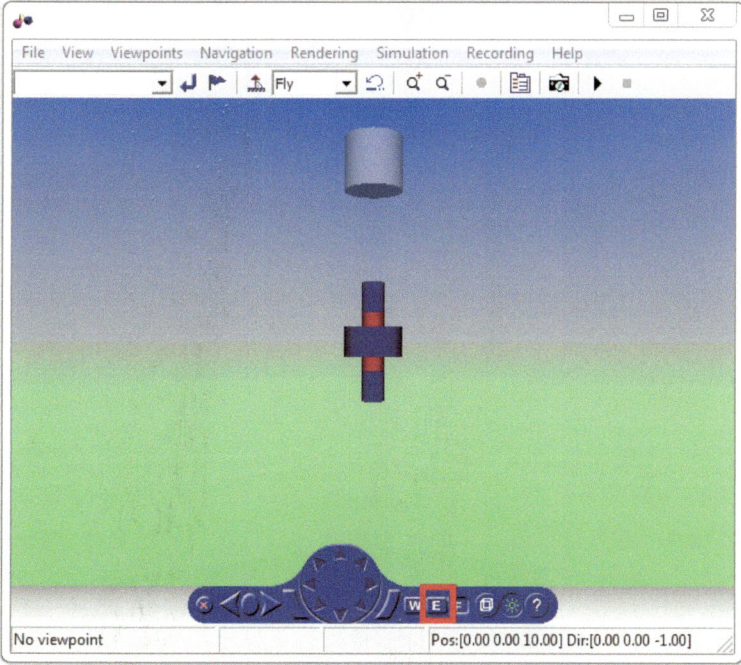

Fig. 10.17 Examine button

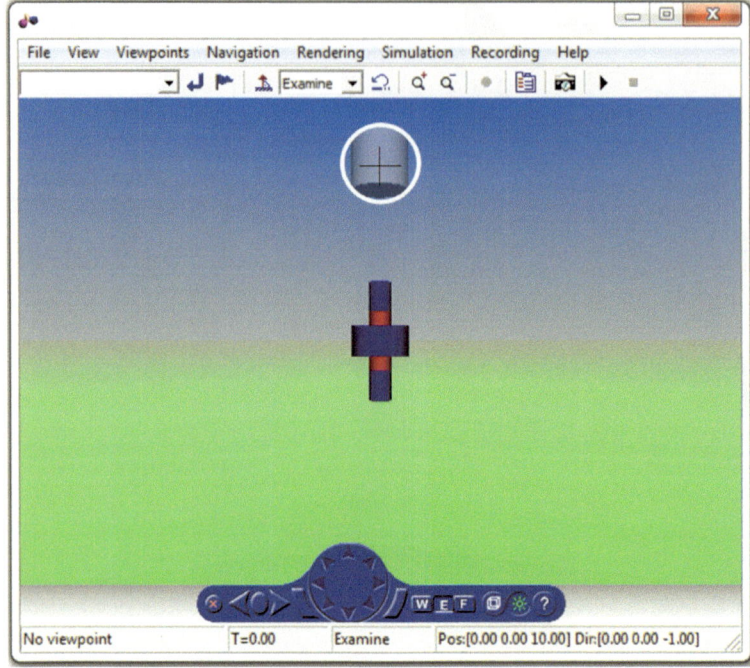

Fig. 10.18 The cursor changes to a cross

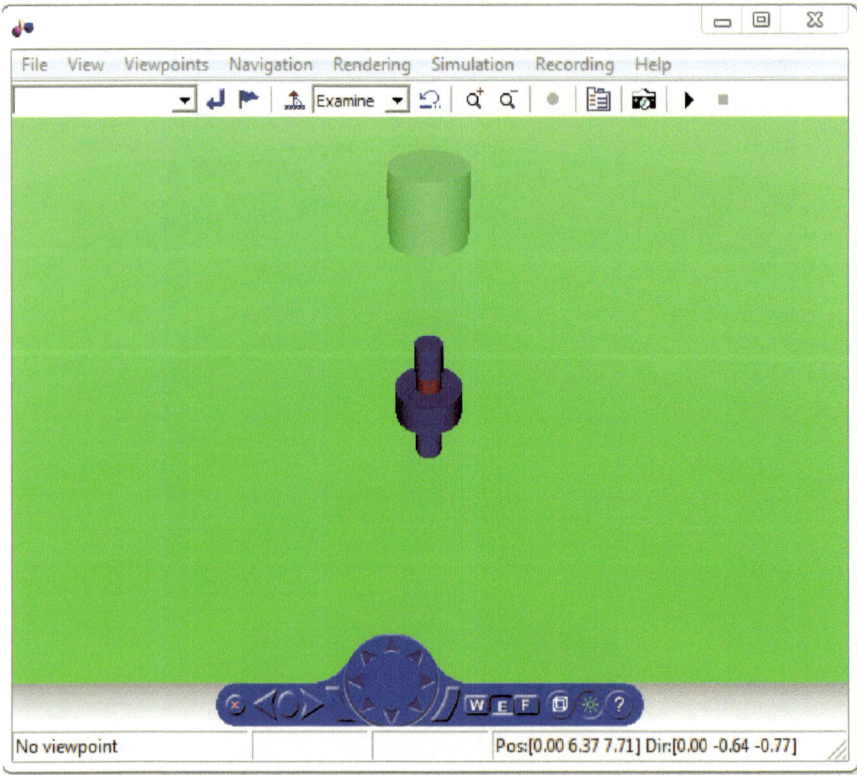

Fig. 10.19 The rotated view

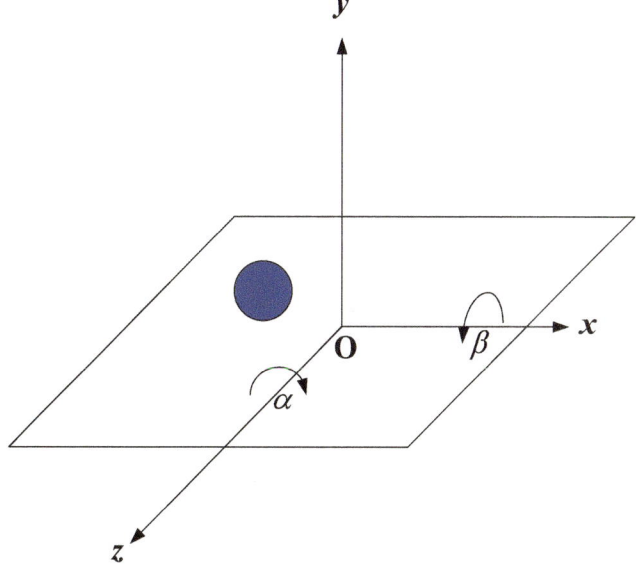

Fig. 10.20 Ball on a plate in addition to the rotation angles of the plate

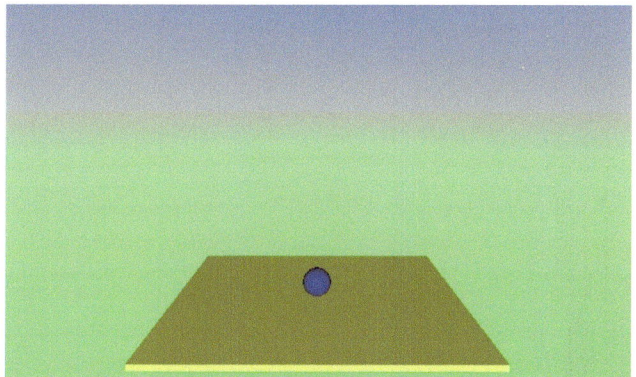

Fig. 10.21 Virtual scene of the ball and the plate

References

Books

Hibbler RC (1986) Engineering mechanics: dynamics, 4th edn. McMillan, New York
Ogata K (2010) Modern control engineering, 5th edn. Prentice Hall, New York

Chapter 11
Car Animation with Joystick Control for Simulink® Users

11.1 Introduction

This chapter aims to teach the user how to handle a set of 3D rotations and translations of **VRML** bodies. Furthermore, the user will learn how to utilize human interface devices (such as the joystick) to control and interact with the **VRML** bodies through a Simulink® model.

There are many ways to describe 3D rotations of bodies. For example:

1. Euler rotations (represented by three angular rotations)
2. Rotation vector (represented by a unit vector, $\underline{u}$, and a rotation angle, ϕ)

An Euler rotation around the x_1-axis by an angle ϕ is identical to the rotation vector $\begin{bmatrix} 1 & 0 & 0 & \phi \end{bmatrix}$. This fact will be used to model a series of complex 3D rotations by representing the rotation vector as cascaded Euler rotations. In **VRML** environment, rotations are represented using the rotation vector where the rotation angle is expressed in degrees. Yet the angles from MATLAB® and Simulink® should be in radians when passed to **VRML**. MATLAB® and Simulink® internally handle the conversion of the angles.

In this chapter, the motion of a car in the horizontal plane will be animated. The steering of the car and the traction force will be controlled by means of a joystick.

In Sect. 11.2, the governing equations of motion of the car in the horizontal plane will be introduced. The car is subject to a steering action and traction force. The various degrees of freedom of the system will be defined along with the discrete empirical equations of motion.

Section 11.3 will guide the reader to set up the joystick that will be used to control the car.

In Sect. 11.4, the Simulink® model that will use the control commands from the joystick as an input to animate the car will be developed.

Matlab® and Simulink® are registered trademarks of The Mathworks, Inc.

N. Khaled, *Virtual Reality and Animation for MATLAB® and Simulink® Users: Visualization of Dynamic Models and Control Simulations*, DOI 10.1007/978-1-4471-2330-9_11, © Springer-Verlag London Limited 2012

The chapter concludes by a fuzzy logic application for the speed control of the car.

The electronic version of all the Simulink® models, **VRML** models, recorded movies for the animated car, in addition to the fuzzy controller for the speed can be downloaded from Springer's web site http://extras.springer.com/.

11.2   Equations of Motion of the Car and Wheels

Given a car that is moving in the xz-plane. The two inputs that are responsible for the planar motion of the car are the traction force F and the steering angle, θ, of the front wheels (the road will be assumed horizontal in the xz-plane). These two inputs are modified by the user instantaneously through the joystick. The inertial coordinate system $\{x, y, z\}$ is centered at O (in **VRML**, this system has the x- and y-axes to be the horizontal and vertical directions of the computer's screen, respectively). The first body-fixed coordinate system $\{x_1, y_1, z_1\}$ is attached to the car and centered at its center, C (Fig. 11.1), and is fixed to the car. The angle between the x-axis and the x_1-axis is β (Fig. 11.1). Two body-fixed coordinate systems $\{x_2, y_2, z_2\}$ of center W_{f_l} and $\{x_3, y_3, z_3\}$ of center W_{f_r} are used to describe the orientation and position of the front left and right wheels, respectively. Since both front wheels have the same steering angle θ, the two coordinate systems $\{x_2, y_2, z_2\}$ and $\{x_3, y_3, z_3\}$ have the same unit vectors. The rear left and right wheels are represented by their centers W_{r_l} and W_{r_r} in Fig. 11.1. The degrees of freedom that will be used to model the car and wheels are the horizontal coordinates of point C (X_C and Z_C), orientation of the car, β, in addition to the steering angle of the front wheels, θ.

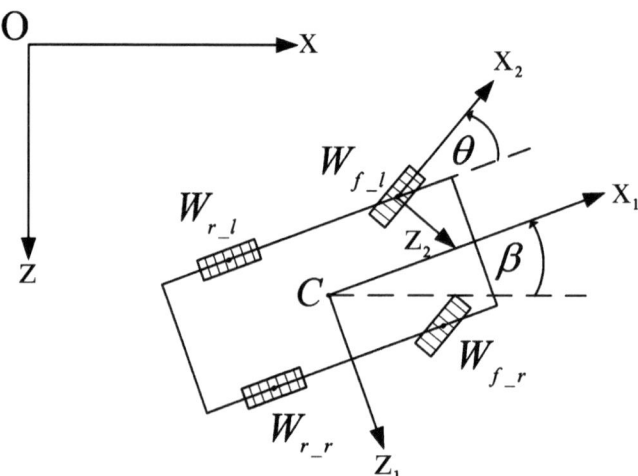

Fig. 11.1 Rotation angles for the car and the steering angle

The velocity of the car in the x_1 direction is V. The latter is denoted by $V(i)$ and $V(i+1)$ at the previous and current time steps, respectively. $V(i+1)$ is proportional to $V(i)$ and the current traction force $F(i+1)$, and it is computed using the following empirical formula:

$$V(i+1) = 0.3 \times V(i) + F(i+1) \tag{11.1}$$

The velocity of the car, $V(i+1)$, is saturated at its maximum forward value 3.4, and −0.2 m/s in the reverse direction.

The inertial velocity of the car is computed using the rotation matrix between the coordinate systems (x_1, y_1, z_1) and (x, y, z):

$$\begin{bmatrix} V_x(i+1) \\ V_y(i+1) \\ V_z(i+1) \end{bmatrix} = \begin{bmatrix} \cos(\beta(i)) & 0 & \sin(\beta(i)) \\ 0 & 1 & 0 \\ -\sin(\beta(i)) & 0 & \cos(\beta(i)) \end{bmatrix} \begin{bmatrix} V(i) \\ 0 \\ 0 \end{bmatrix} \tag{11.2}$$

Notice that $V_Y(i+1) = 0$ due to the planar motion of the car.

Using the inertial velocity from Eq. 11.2, one can compute the inertial position of the car by computing the discrete integral of the inertial velocity:

$$\begin{bmatrix} x(i+1) \\ y(i+1) \\ z(i+1) \end{bmatrix} = \begin{bmatrix} x(i) \\ y(i) \\ z(i) \end{bmatrix} + \Delta t \times \begin{bmatrix} V_x(i+1) \\ V_y(i+1) \\ V_z(i+1) \end{bmatrix} \tag{11.3}$$

where Δt is the step size.

The orientation of the car, β, is a function of the steering angle θ and the velocity V. $\beta(i+1)$ is computed based on the following empirical formula:

$$\beta(i+1) = \beta(i) + 0.05 \times \theta(i+1) \times V(i+1) + 0.9 \times (\theta(i+1) - \theta(i)) \times V(i+1) \tag{11.4}$$

The current traction force, $F(i+1)$, is computed as follows:

$$F(i+1) = F(i) - 0.0005 \times round(joystick\_up\_down) \tag{11.5}$$

where $round(joystick\_up\_down)$ is equal to +1 or −1 if the up or down button of the joystick is pressed, respectively.

The current steering angle, $\theta(i+1)$, is computed as follows:

$$\theta(i+1) = \theta(i) - 0.001 \times round(joystick\_left\_right) \tag{11.6}$$

where $round(joystick\_left\_right)$ is equal to +1 or −1 if the right or left button of the joystick is pressed, respectively. The lower and upper limits of the steering angle of the car, $\theta(i+1)$, are $\dfrac{-\pi}{4}$ and $\dfrac{\pi}{4}$, respectively.

Figure 11.2 shows a hierarchy view of the virtual scene of the car, *car.wrl*, that has been created in Chap. 6. The first level is the virtual scene which is created

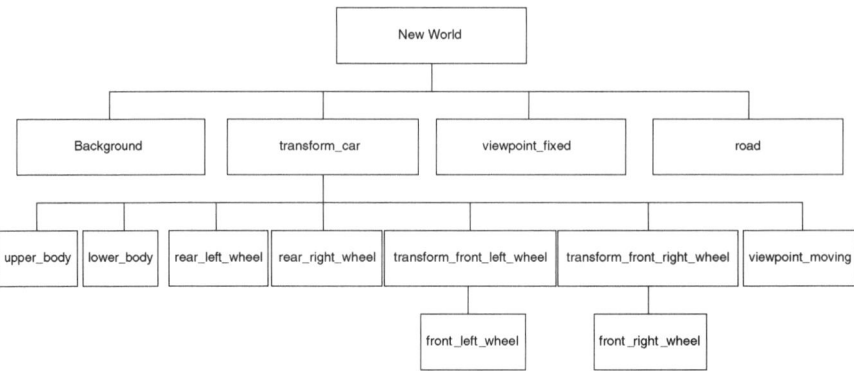

Fig. 11.2 Hierarchy view of the virtual scene of the car

automatically with each new project. On the second level of the virtual scene, there are four components:

1. **The Background**: It is the background of the virtual scene.
2. **The transform_car**: It is the **Transform** that will be manipulated in the Simulink® model to translate and rotate the car, and thus the **Children** (components) of **transform_car** will translate and rotate accordingly.
3. **The viewpoint_fixed**: It is a fixed **Viewpoint**.
4. **The road:** It is a straight road. It is a **Box** component.

On the third level of the virtual scene, there are seven **Children** of **transform_car.** The **upper_body** and **lower_body** are **Box** components, and they form the body of the car. The **rear_left_wheel** and **rear_right_wheel** are the rear wheels of the car, and they are **Cylinder** components. **The transform_front_left_wheel** and **transform_front_right_wheel** are two **Transform** components that are used as additional layers to rotate the left and right wheels, respectively, using the steering angle (these two **Transforms** are optional, and the user could rotate the front wheels without adding them, but for complex systems, transformations become cumbersome; thereby the author advises the user to add the extra **Transforms**). Finally, **viewpoint_moving** is a **Viewpoint** that is moving with the car.

11.3   Joystick Setup

Any simple USB joystick could be used along with this example. The one that has been used to run this example is "Logitech Precision Joystick" (Fig. 11.3). It does not require any setup. It should be a plug-and-play device, and it is recommended to be plugged in the USB port before starting the MATLAB® session. The buttons that will be used in running this example are shown in Fig. 11.3.

To test and identify the USB joystick, follow the steps below:

1. Connect the USB joystick.
2. Start MATLAB® and open a new Simulink® model.

Fig. 11.3 "Logitech Precision Joystick" with the required buttons for this chapter

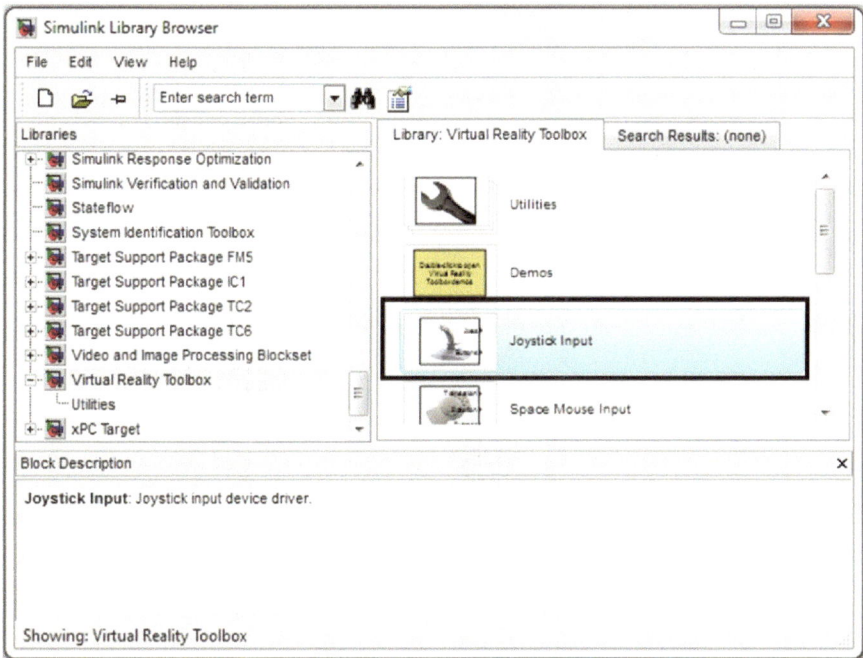

Fig. 11.4 Joystick Input block in Virtual Reality Toolbox

3. Insert a Joystick Input block from the **Virtual Reality Toolbox** (Fig. 11.4).
4. The **Joystick Input** block has two outputs: **Axes** (up-down-left-right buttons) and **Buttons** (the rest of the buttons). Connect the Axes output to a **Display** sink.

5. Change the **Stop time** of the Simulink® model to **inf**. Run the Simulink® model and press the up button, down button, left button, and right button, respectively, and denote the values displayed by the **Display** sink. The values that the author got by running this simple system-identification experiment were −0.9922, 1, −0.9922, and 1, respectively. These values might be slightly different for various joysticks and/or computers, yet the round of these numbers should be the same (either −1 or 1).
6. Stop the model and close it.

11.4   Creating the Simulink® Model for Animating the Virtual Scene

In what follows, the virtual scene, *car.wrl*, will be animated by varying the traction force and the steering angle by using the joystick. The variables that will be changed in the virtual world are:

1. **The translation** property of **transform_car** (which represents the translation of the car with the wheels)
2. **The rotation** property of **front_right_wheel** (which represents the rotation of the front right wheel by the steering angle, θ)
3. **The rotation** property of **front_left_wheel** (which represents the rotation of the front left wheel by the steering angle, θ)

Follow the steps given below to create the Simulink® model that will animate the virtual scene (assuming that the user has the **Virtual Reality Toolbox*** installed in Simulink®):

1. Create a new Simulink® model.
2. Add a **VR Sink** from the **Virtual Reality Toolbox*** to the current model (Fig. 11.5). This sink will contain the virtual scene, *car.wrl*, and will allow the user to manipulate the scene through Simulink® inputs.
3. Double-click the **VR Sink** in the model. A window for the parameters of the VR Sink will open (Fig. 11.6). Click on the **Browse** button (Fig. 11.6) to navigate and select the virtual scene *car.wrl* which should be in Chap. 11 material of the book. Download the file to the working directory of MATLAB®. Click **Open** once you select the file *car.wrl*.
4. Keep the **Sample time** of the **VR Sink** 0.1 (Fig. 11.7). Note that **Sample time** should be equal to (or a multiple of) the one set in the Simulink® model.
5. Click on the + sign to the left of **transform_car (Transform)** (Fig. 11.8) to expand its elements.
6. Tick the squares to the left of **rotation** and **translation** properties of **transform_car** (Fig. 11.9). This will enable the user to manipulate the translation of the car through Simulink® inputs. Note that these two properties are represented in the inertial frame as they do not belong to any other **Transform** in the virtual scene.

---

*As of Matlab® 2009a, **Virtual Reality Toolbox** has been renamed **Simulink 3D Animation**.

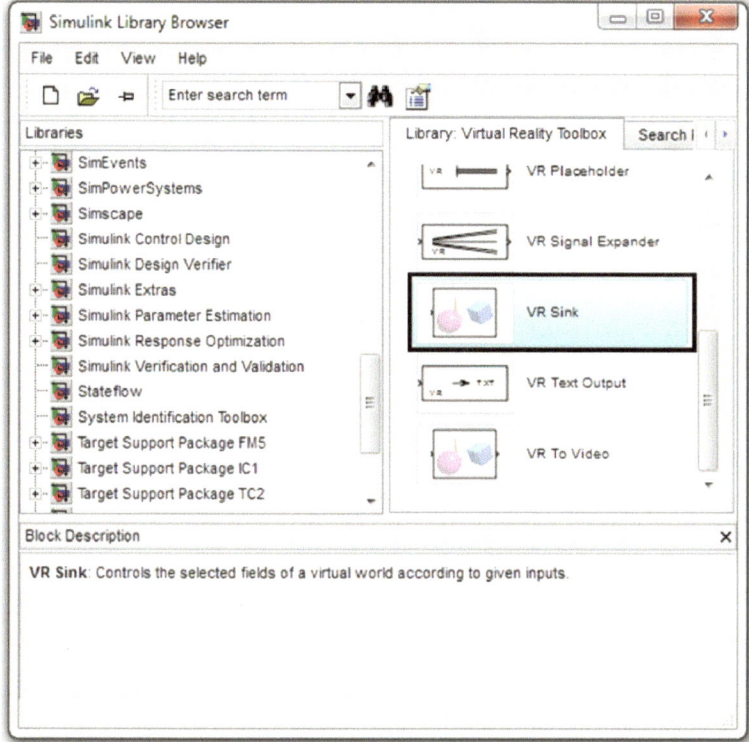

Fig. 11.5   VR Sink in Virtual Reality Toolbox*

7. Click the + sign on the left of **Children** to expand its elements (Fig. 11.10).
8. Click the + sign on the left of **transform_front_left_wheel** to expand its elements (Fig. 11.11).
9. Tick the square to the left of **rotation** property of **transform_front_left_wheel** (Fig. 11.12). Note that the **rotation** property will enable the user to manipulate the rotation of the left wheel in the local frame with respect to its axis. Refer to Chap. 6 to better understand the structure of the virtual scene.
10. Repeat steps 8 and 9 for **transform_front_right_wheel.** Click **OK** when done.
11. Figure 11.13 shows the **VR Sink** for *car.wrl* with all the inputs to the sink properly setup.
12. In the Simulink® model, go to **Simulation>Configuration Parameters** and change the solver type to **Fixed-step** (Fig. 11.14). Note that the **step size** should be set to 0.1 to match the **Sample time** of the **VR Sink**. In addition, change the **Stop time** to inf. Click **OK** when done.

---

*As of Matlab® 2009a, **Virtual Reality Toolbox** has been renamed **Simulink 3D Animation**.

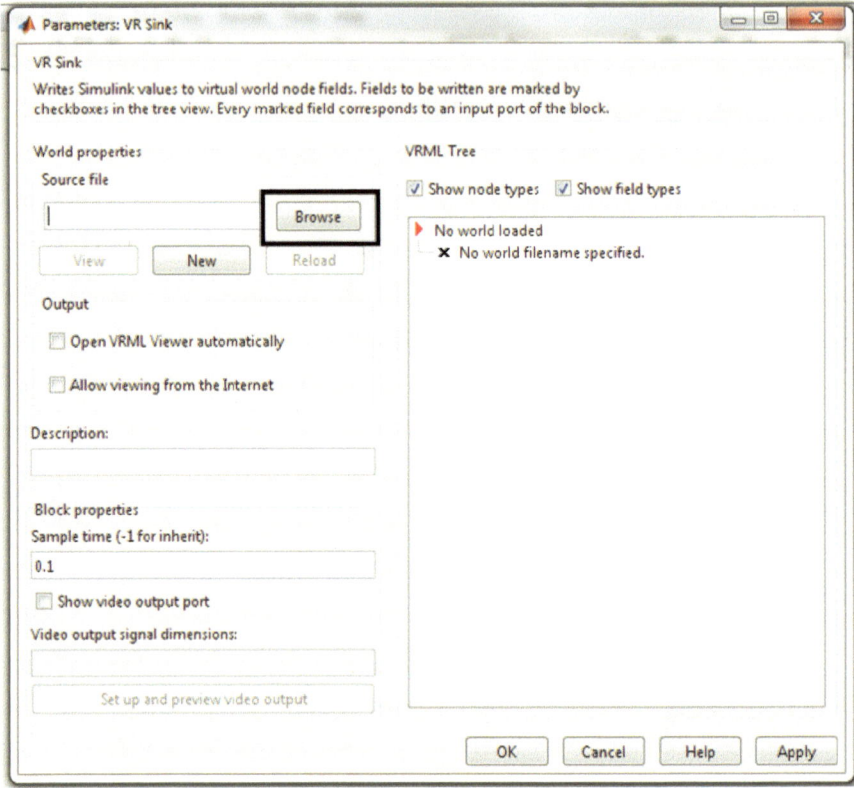

Fig. 11.6 Browse button for the virtual scene is shown in the rectangle

13. Insert a **Joystick Input** block from the **Virtual Reality Toolbox** (Fig. 11.4).
    Figure 11.15 shows the **Joystick Input** block connected to the terms
    $round(joystick\_up\_down)$ and $round(joystick\_left\_right)$ used in Eqs. 11.5
    and 11.6.
14. Figure 11.16 shows the solution of Eqs. 11.5 and 11.6 with the saturation of
    the steering angle.
15. Figure 11.17 shows the solution of Eqs. 11.1 and 11.4 with the saturation of
    the velocity.
16. Figure 11.18 shows the solution of Eq. 11.2 to transform the velocity to the
    inertial frame.
17. Figure 11.19 shows the solution of Eq. 11.3 to compute the inertial position of
    the car.
18. Figure 11.20 shows the inputs that are fed to the **VR Sink**.
19. Save the model.

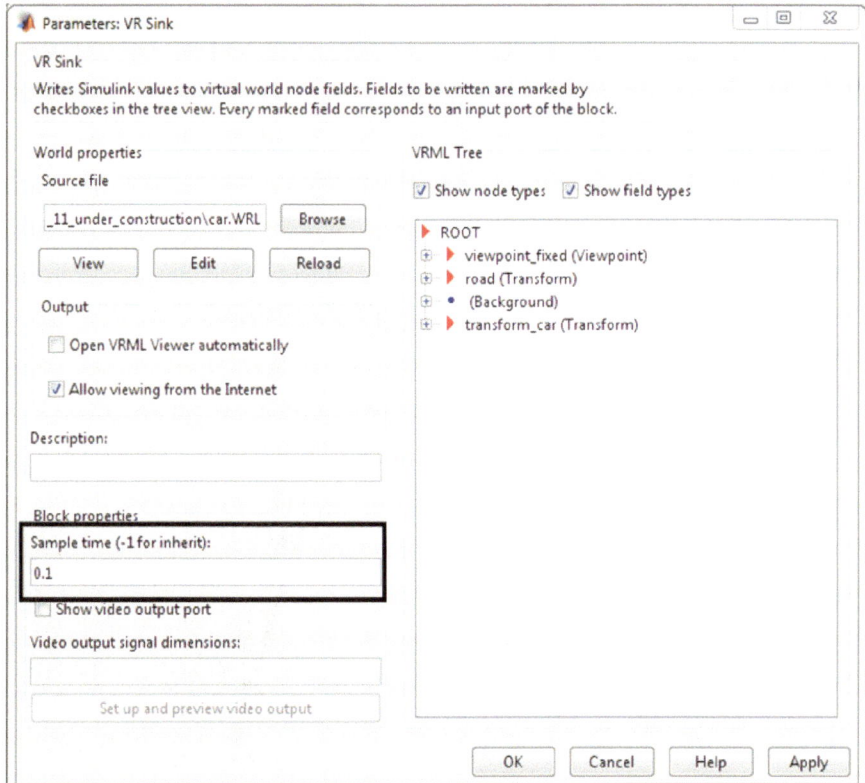

Fig. 11.7 Sample time for VR Sink

20. Double-click the **VR Sink** to open the virtual scene. Then run the model and use the USB joystick to increase the speed and steer the car.

The user can change the USB joystick gains used in the Eqs. 11.5 and 11.6 to increase/decrease the amplitude of the steering angle and the thrust commanded by the joystick. The solution of the problem can be downloaded from Springer's web site http://extras.springer.com/. The files can be found in folder /Chapter 11.

11.5   Application Problem: Fuzzy Logic Controller for the Car Speed

The main objective of this section is to design a speed controller for the car model presented in Sect. 11.2. In this problem, the car steering angle will still be controlled by the joystick, but the traction force, $F(i+1)$, will be commanded by the controller

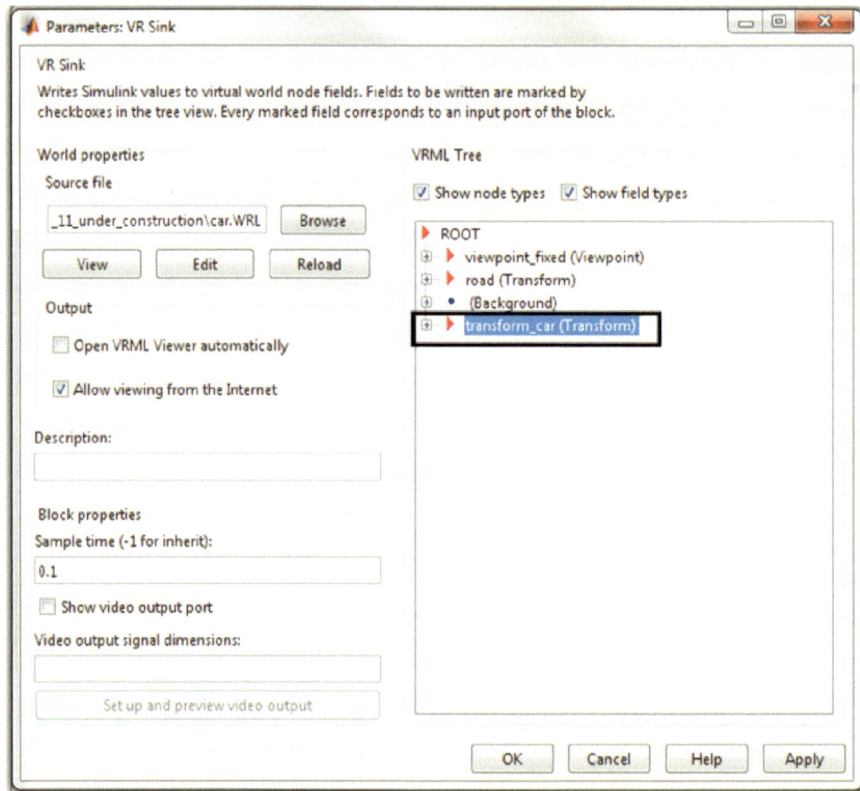

Fig. 11.8   The transform_car (Transform)

to maintain a fixed speed of 3 m/s for the car. The car should reach the desired speed in a maximum time of 3 s with zero overshoot. The fuzzy logic controller (Yen and Langari 1999) is incremental (i.e., the value of $F(i+1) = F(i) + \delta F(i)$, where δF is the current output of the fuzzy logic controller). The structure of the fuzzy logic controller is of the following form (Fig. 11.21): e is the speed error and is defined as $V_{t\,arg\,et}(i) - V(i) = 3 - V(i)$. K_e, K_{int-e}, and K_F are the gains of the controller. The 25 rules for the fuzzy logic controller are summarized in Table 11.1.

Figures 11.22, 11.23, and 11.24 show the plots for the membership functions for $e(i)$, $\int_{0}^{t} e(i)d\tau$, and $\delta F(i)$. Solve the following:

1. Using the MATLAB® Fuzzy Logic Toolbox (type in **fuzzy** in the MATLAB® workspace), construct the speed fuzzy controller based on the rules given in Table 6.1 and the membership functions in Figs. 11.22, 11.23, and 11.24.

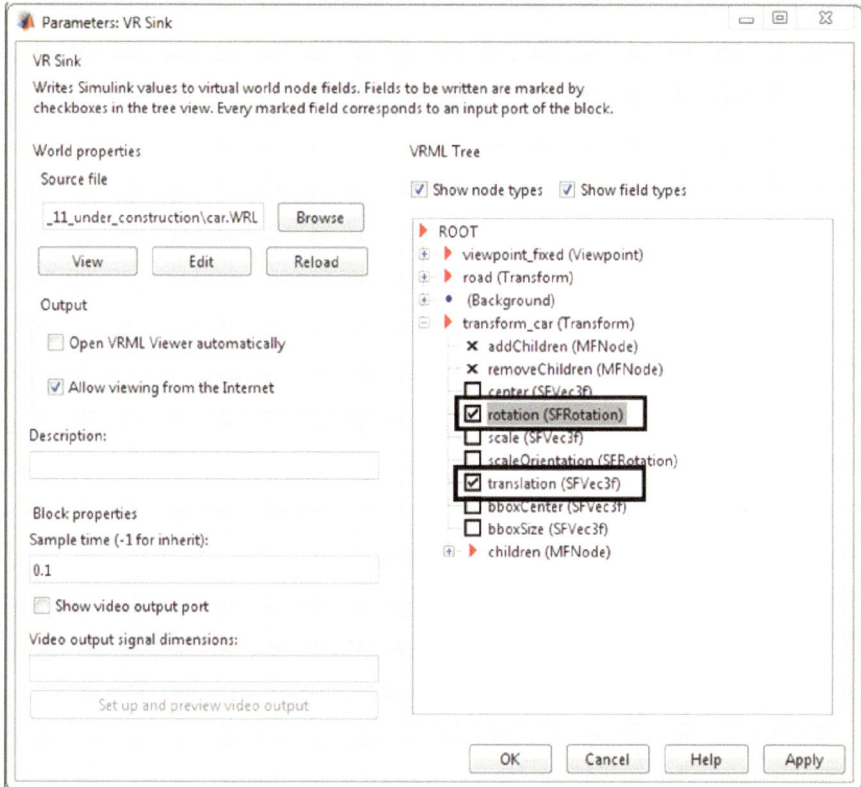

Fig. 11.9 The rotation and translation property of transform_car

2. Use the developed controller along with the Simulink® model developed to animate the car (***car_animate.m***) in order to animate the controlled speed of the car in VRML. Plot the instantaneous and desired speeds of the car while running the simulation.

3. Tune the controller by playing around with the gains K_e, K_{int-e}, and K_F in order to meet the requirements.

The solution of the problem can be downloaded from Springer's web site http://extras.springer.com/. The files can be found in folder /Chapter 11/Application Problem.

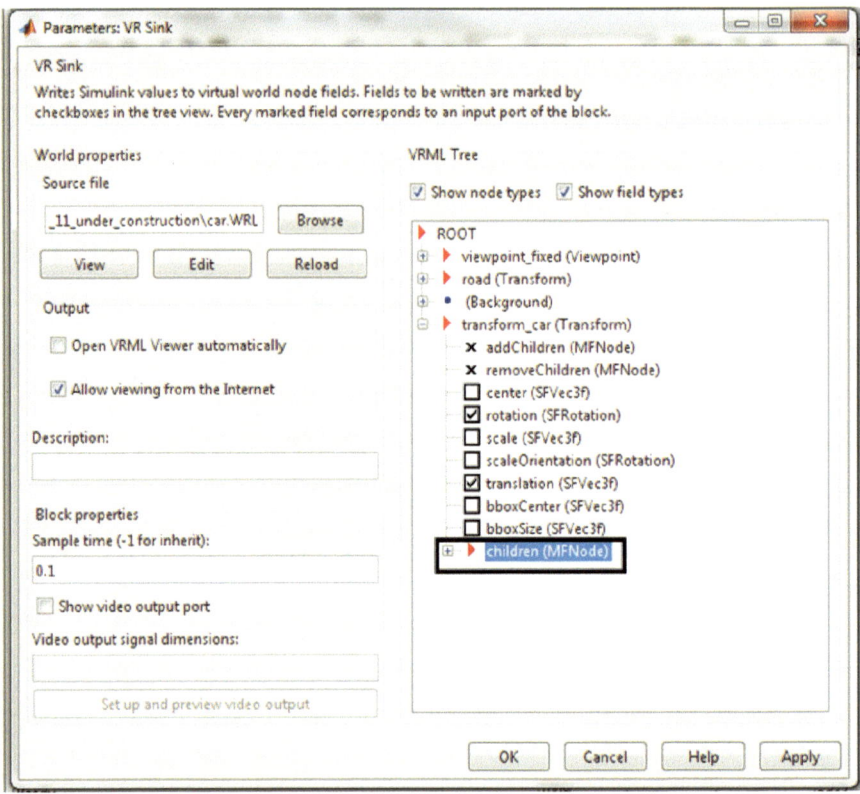

Fig. 11.10 Children of transform_car

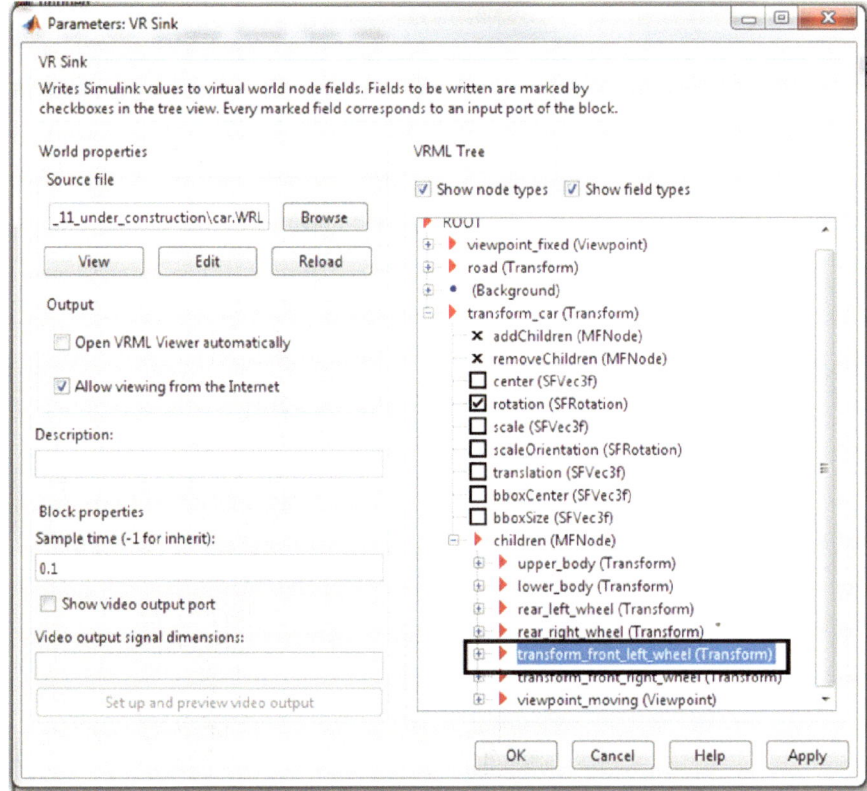

Fig. 11.11 The transform_front_left_wheel

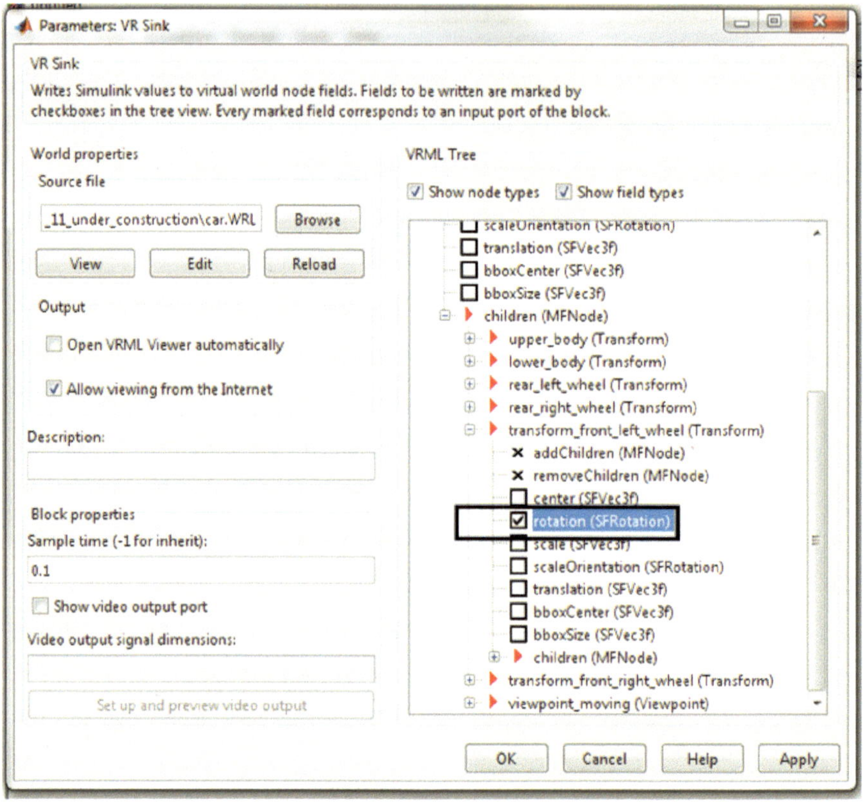

Fig. 11.12 The rotation property of transform_front_left_wheel

Fig. 11.13 VR Sink
for car.wrl

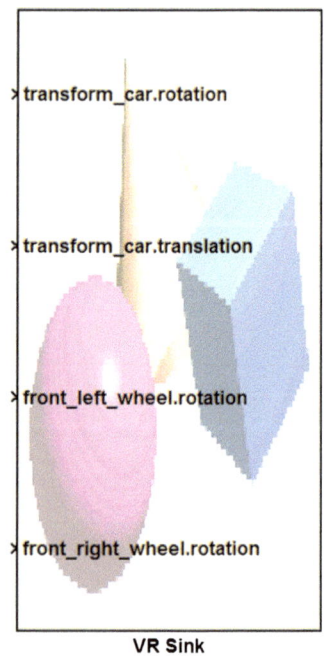

VR Sink

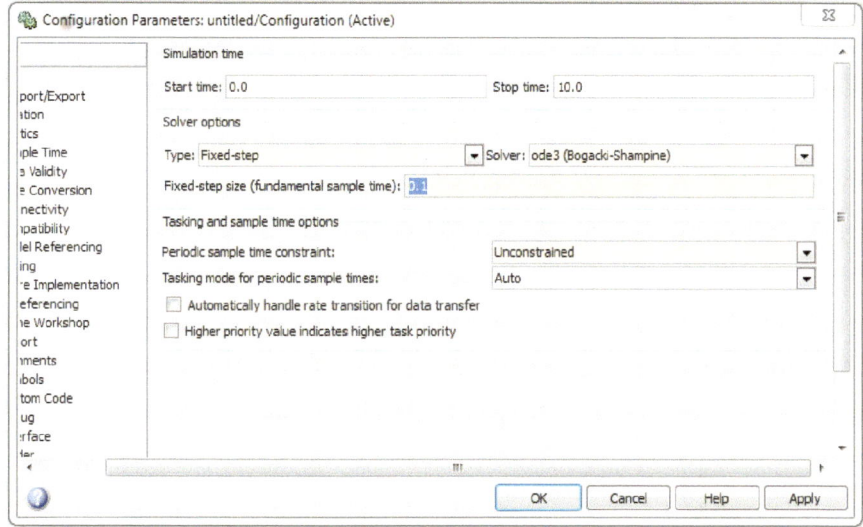

Fig. 11.14 Solver options

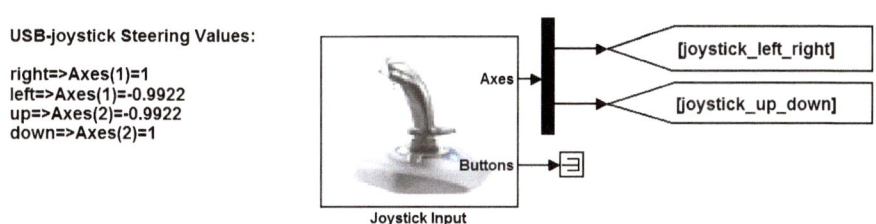

Fig. 11.15 Connections of the Joystick Input block

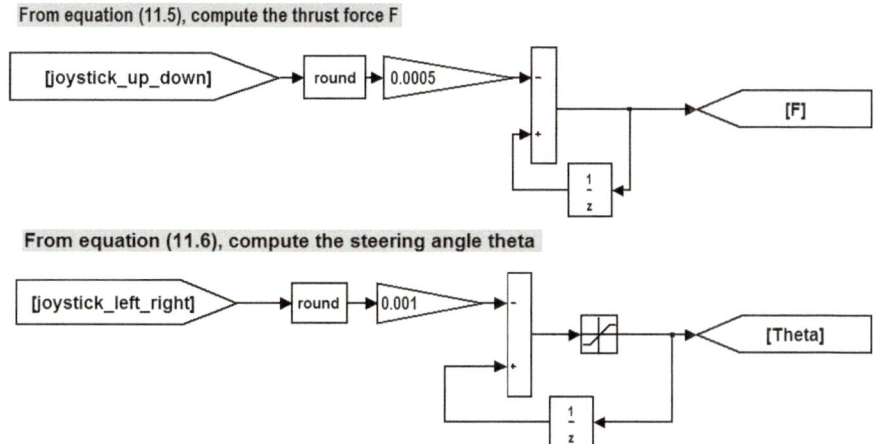

Fig. 11.16 Solving Eqs. 11.5 and 11.6 with the saturation of the steering angle

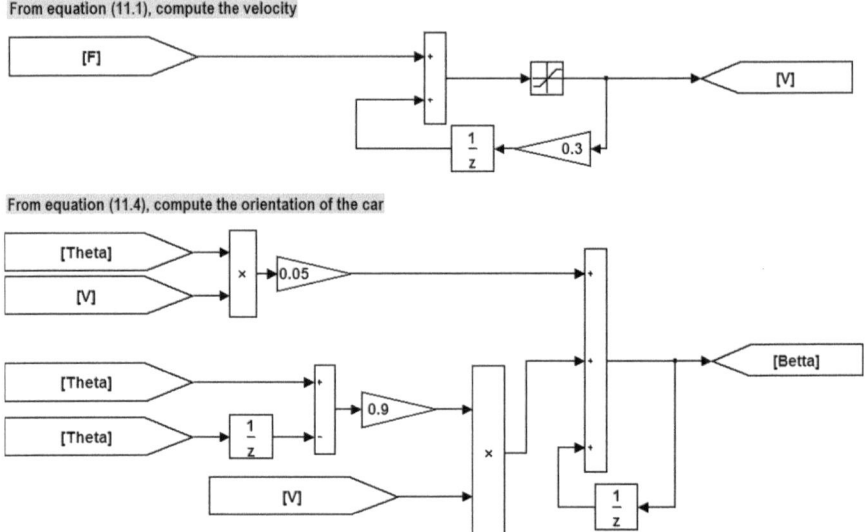

Fig. 11.17 Solving Eqs. 11.1 and 11.4 with the saturation of the velocity

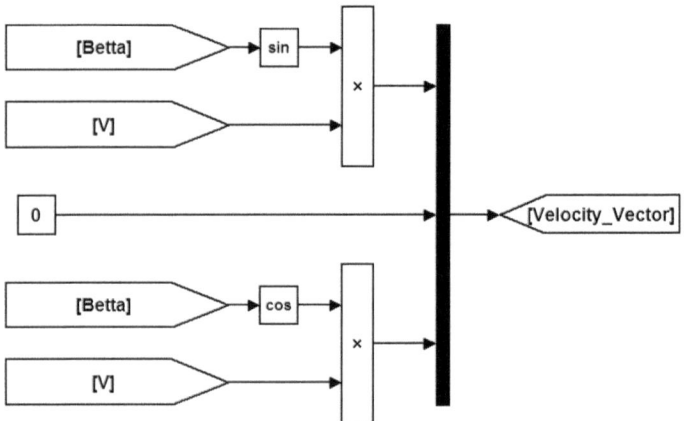

Fig. 11.18 Solving Eq. 11.2 to transform the velocity to the inertial frame

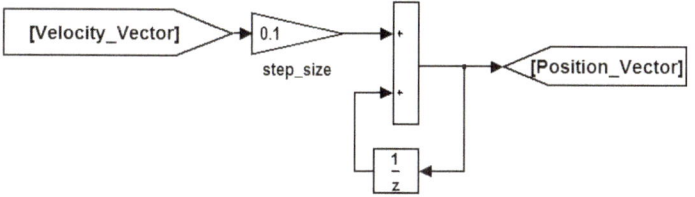

Fig. 11.19 Solving Eq. 11.3 to compute the inertial position of the car

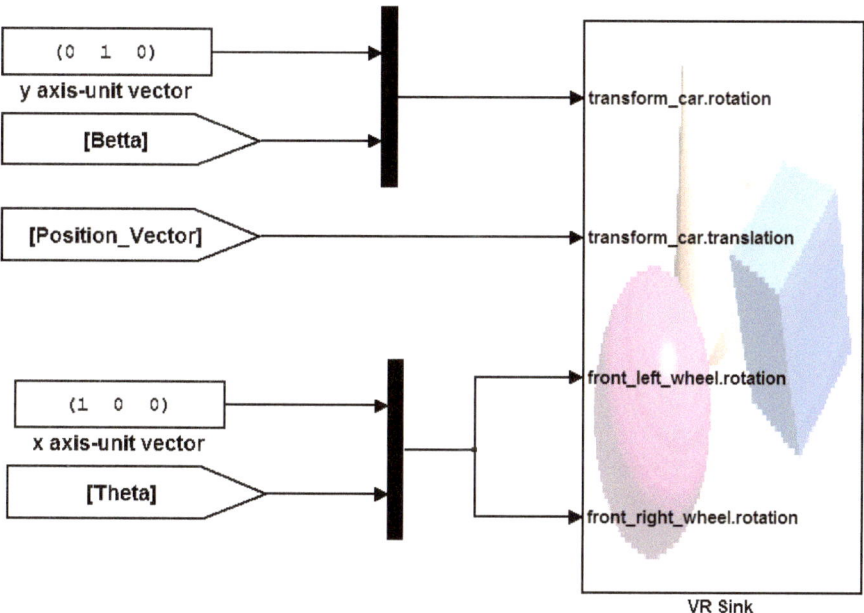

Fig. 11.20 Feeding the inputs to the VR Sink

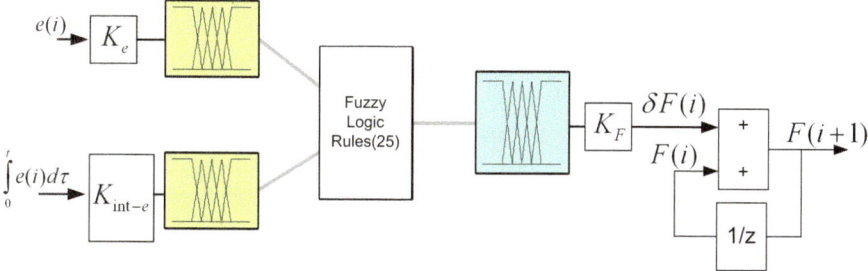

Fig. 11.21 Incremental fuzzy logic speed controller

Table 11.1 Rules for the
incremental fuzzy logic
controller

	e				
$\int e$	NL	NS	Z	PS	PL
NL	NL	NL	NS	PS	PL
NS	NL	NS	Z	PS	PL
Z	NL	NS	Z	PS	PL
PS	NL	NS	Z	PS	PL
PL	NL	NS	PS	PL	PL

NL stands for negative large, *NS* stands
for negative small, *Z* stands for zero,
PS stands for positive small and *PL*
stands for positive large

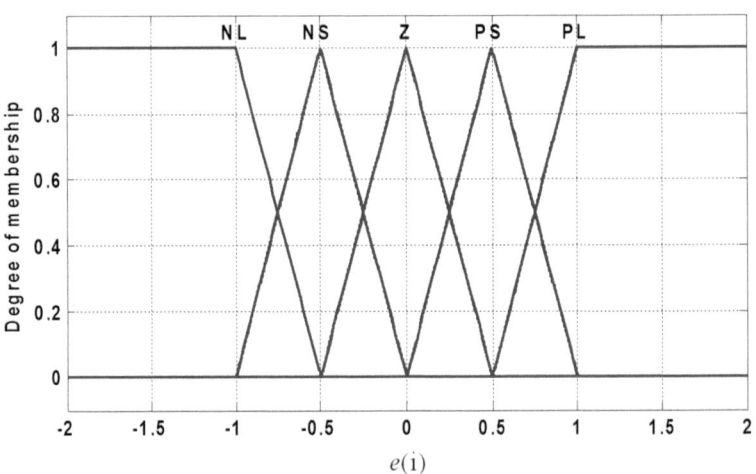

Fig. 11.22 Membership functions for the error $e(i)$

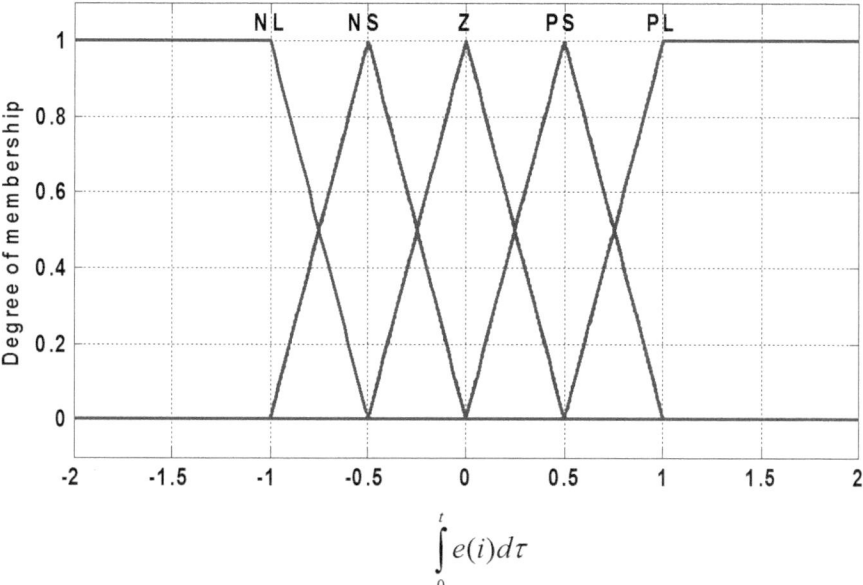

Fig. 11.23 Membership functions for the integral of the error $\int_{0}^{t} e(i)\mathrm{d}\tau$

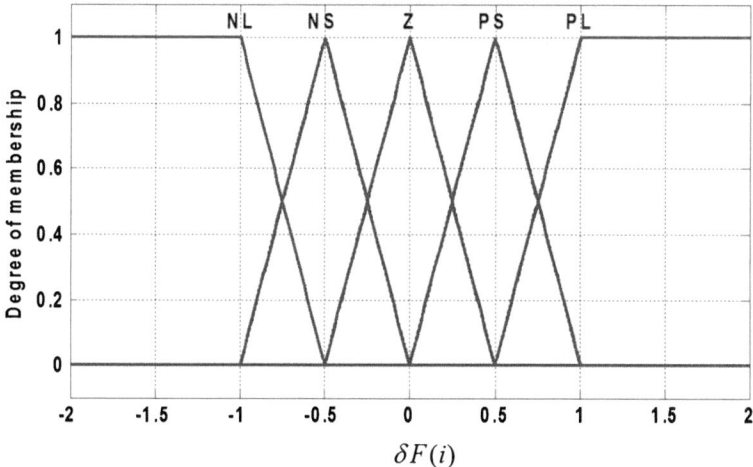

Fig. 11.24 Membership functions for the increment of the controller output

Reference

Book

Yen J, Langari R (1999) Fuzzy logic: intelligence, control and information. Prentice Hall, Upper
    Saddle River

Chapter 12
Animation of a Ship Moving Across Waves Using Simulink®

12.1 Introduction

This chapter aims to teach the reader how to handle meshing and animation of a 3D surface such as the sea surface using Simulink®. The virtual scene, *sea_scene.wrl*, that was created in Chap. 7 included sea waves in addition to a ship (Fig. 12.1). In this chapter, the sea waves and the motion of the ship will be animated in Simulink®.

In Sect. 12.2, the governing equations of motion of the waves and the ship will be introduced.

In Sect. 12.3, the Simulink® that will be used to animate the scene will be introduced.

The chapter concludes by an application problem for controlling the ship heading. The problem will include the development and tuning of a fuzzy logic controller for the heading of the ship. It will also include the creation of a Simulink® model to animate the virtual scene, *sea_scene2.wrl*, which was created in Chap. 7.

The electronic version of all the Simulink® models, the virtual scenes used in the chapter, in addition to the fuzzy controller for the heading can be downloaded from Springer's web site http://extras.springer.com/.

12.2 Equations of Motion of Sea Waves and the Ship

In general, sea waves are modeled as short or long crested (Perez 2005 and Khaled 2010). In this chapter, waves will be modeled as a sinusoidal with one frequency ω. These waves will be traveling in the z-direction, and the wave height at a certain point will be computed based on Eq. 12.1:

$$h(t,z) = 2\sin(\omega t + z) \tag{12.1}$$

Matlab® and Simulink® are registered trademarks of The Mathworks, Inc.

N. Khaled, *Virtual Reality and Animation for MATLAB® and Simulink® Users: Visualization of Dynamic Models and Control Simulations*, DOI 10.1007/978-1-4471-2330-9_12, © Springer-Verlag London Limited 2012

Fig. 12.1 Virtual scene of the sea surface, ship, and mountain's background

The ship will be moving along the x-direction, starting from the initial position $x(0) = -100$, according to Eq. 12.2:

$$x(t) = -100 + 10t \qquad (12.2)$$

12.3 Simulink® Model for Animating the Virtual Scene

In what follows, the Simulink® model that will animate the virtual scene, ***sea_scene. wrl***, will be developed. The sea surface will vary as a sinusoidal function of time. The ship will move along the x-axis of the virtual scene. The variables that will be changed in the virtual world are:

1. **Height** property of wave_height based on Eq. 12.1
2. **Translation** property of Ship based on Eq. 12.2

Follow the steps below to create the Simulink® model to animate the virtual scene (assuming that the user has the **Virtual Reality Toolbox*** installed in Simulink®):

1. Create a new Simulink® model.
2. Add a **VR Sink** from the **Virtual Reality Toolbox*** to the current model (Fig. 12.2). This sink will contain the virtual scene and will allow the user to manipulate the scene through Simulink® inputs.
3. Double-click the **VR Sink** in the model. A window for the parameters of the **VR Sink** will open (Fig. 12.3). Click on the **Browse** button (Fig. 12.3) to navigate to the virtual scene ***sea_scene.wrl*** which can be downloaded from Chap. 12 material from Springer's web site http://extras.springer.com/. Click Open when you locate the virtual scene.

---

*As of Matlab® 2009a, **Virtual Reality Toolbox** has been renamed **Simulink 3D Animation**.

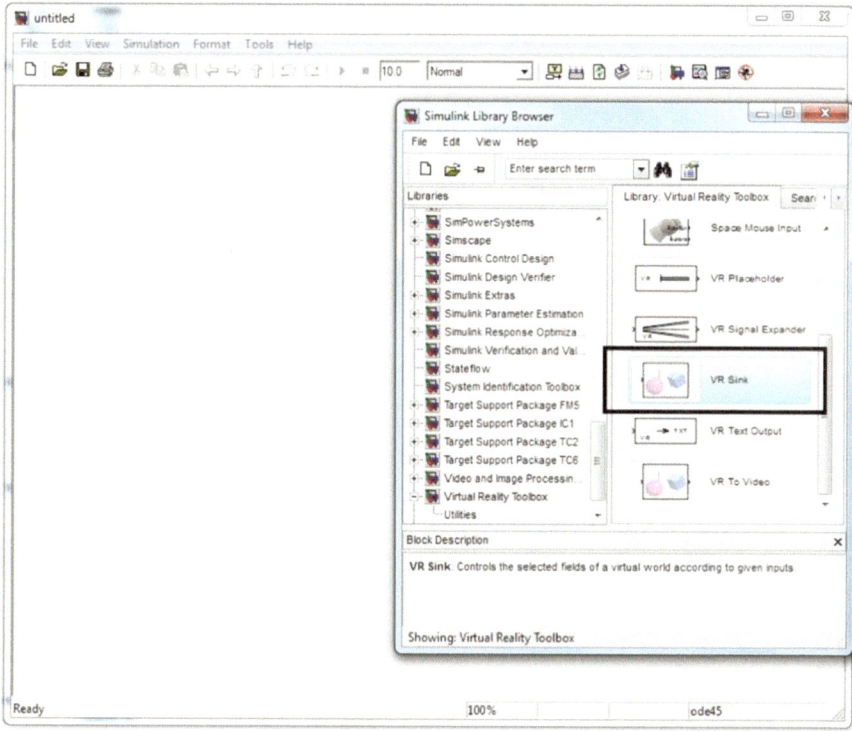

Fig. 12.2 VR Sink in Virtual Reality Toolbox*

4. Change the **Sample time** for the **VR Sink** to 0.01(Fig. 12.4). **Sample time** should match or be a multiple of the one set in the Simulink® model.
5. Click on the + sign to the left of **Ship (Transform)** (Fig. 12.5).
6. Tick the square to the left of translation property to select it (Fig. 12.6).
7. Click on the + sign to the left of **waves (Transform)** (Fig. 12.7).
8. Click on the + sign to the left of **Children** of waves (Fig. 12.8). Similarly, click on the + signs to the left of **(Shape)**, **geometry**, and **wave_height** (Fig. 12.9).
9. Tick the square to the left of the **height** property to select it (Fig. 12.10). Click **OK** to accept the changes and close the parameters of the **VR Sink**.
10. In the Simulink® model, go to **Simulation>Configuration Parameters** and change the solver type to **Fixed-step** and the **step size** to 0.01(Fig. 12.11). Click **OK** when done.
11. Change the Simulation **Stop Time** to 40.
12. Figure 12.12 shows the final model that is used to animate the virtual scene. The user has to be careful about the dimension size of the signals going into **VR Sink**. As for **Ship.translation**, it is a 3×1 signal. Whereas for **wave_height. height**, it is defined by the dimension of the **Elevation Grid** (refer to Chap. 7 for more details).
13. Save the model.

---

*As of Matlab® 2009a, **Virtual Reality Toolbox** has been renamed **Simulink 3D Animation**.

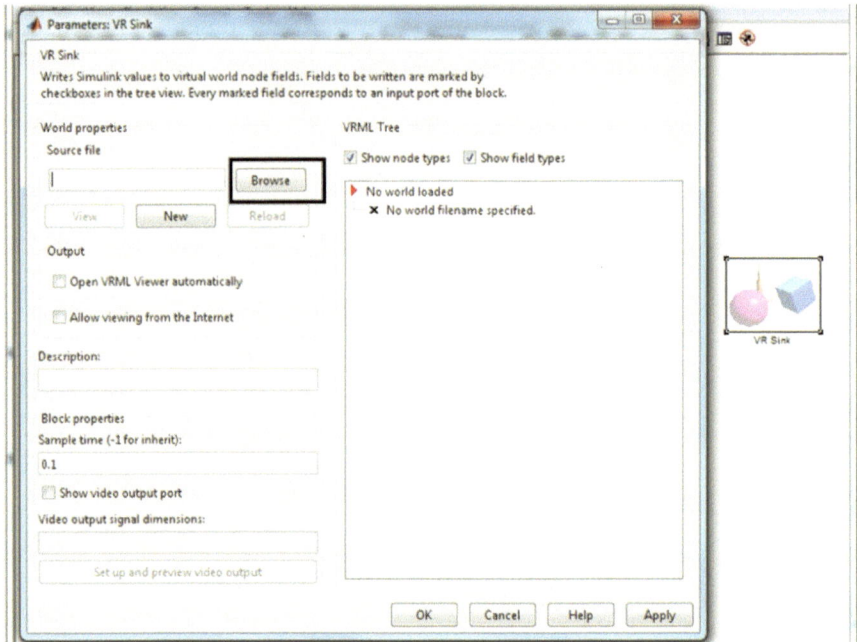

Fig. 12.3 Browse button for the virtual scene is shown in the rectangle

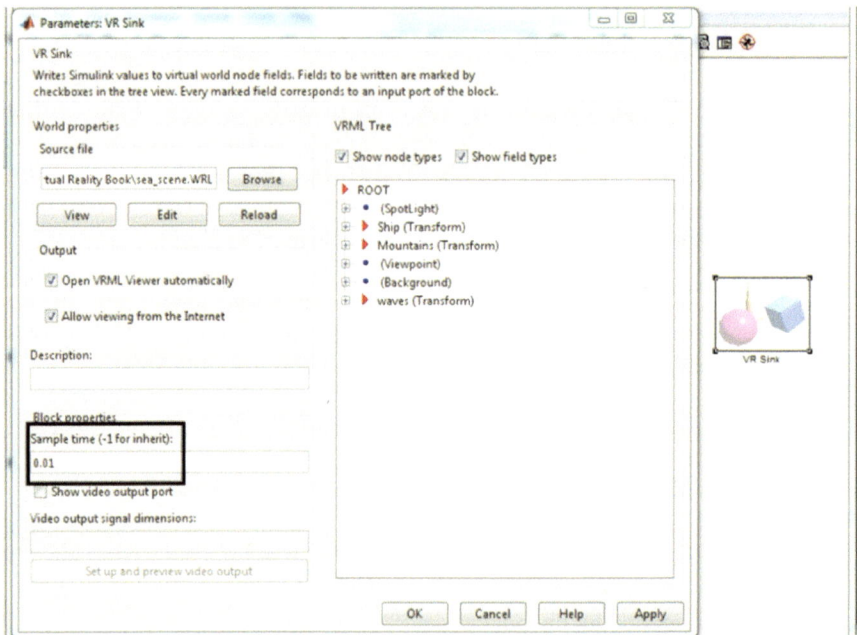

Fig. 12.4 Sample time for VR Sink

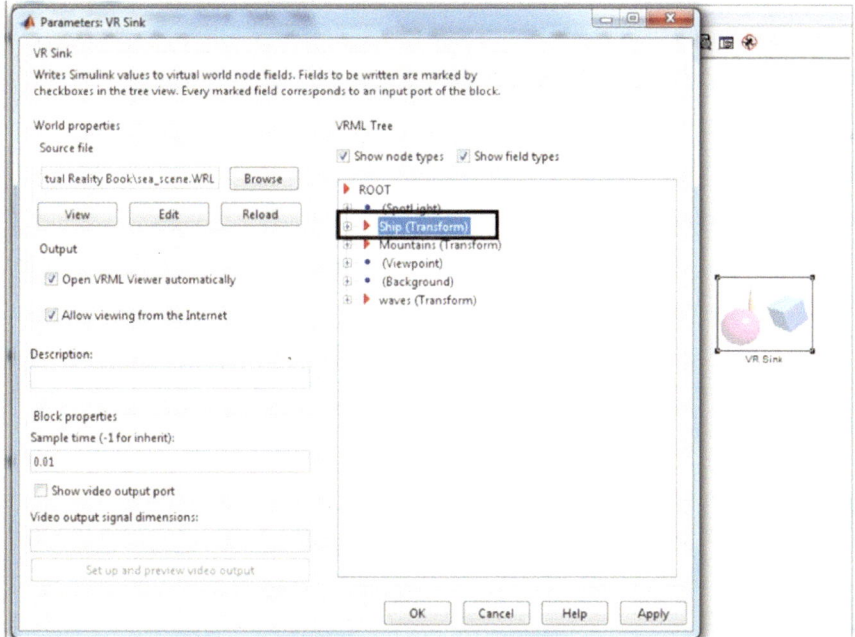

Fig. 12.5 Ship (Transform) is shown in the rectangle

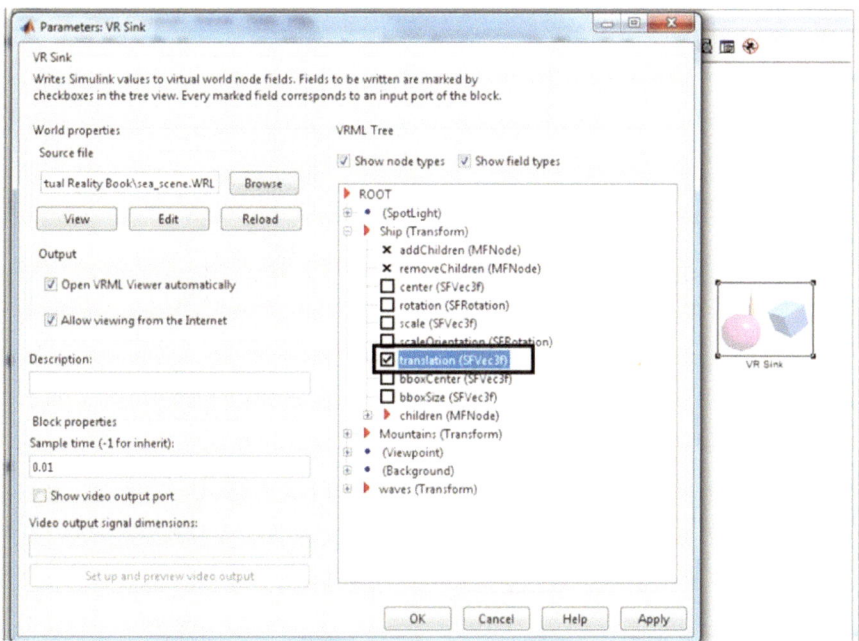

Fig. 12.6 The square to the left of translation property is shown in the rectangle

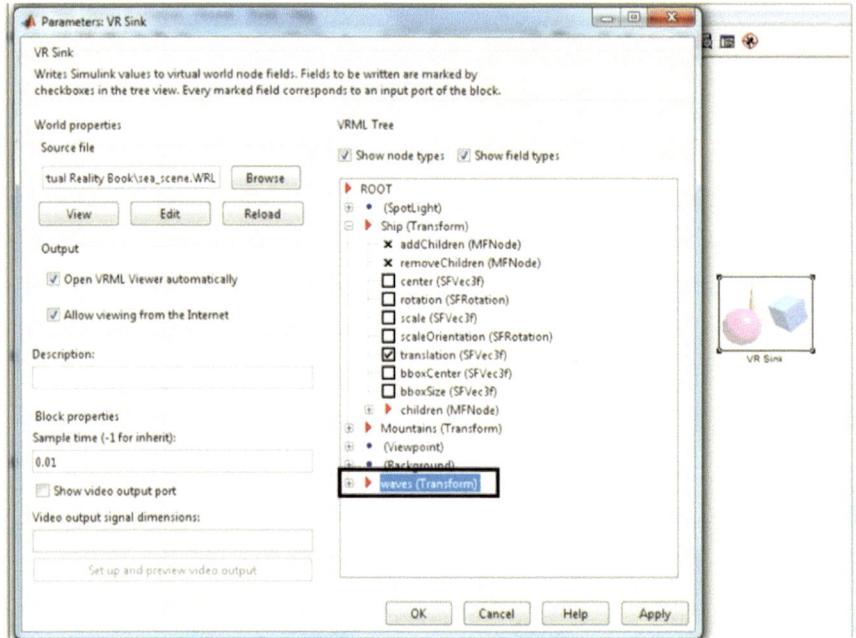

Fig. 12.7   The waves (Transform) is shown in the rectangle

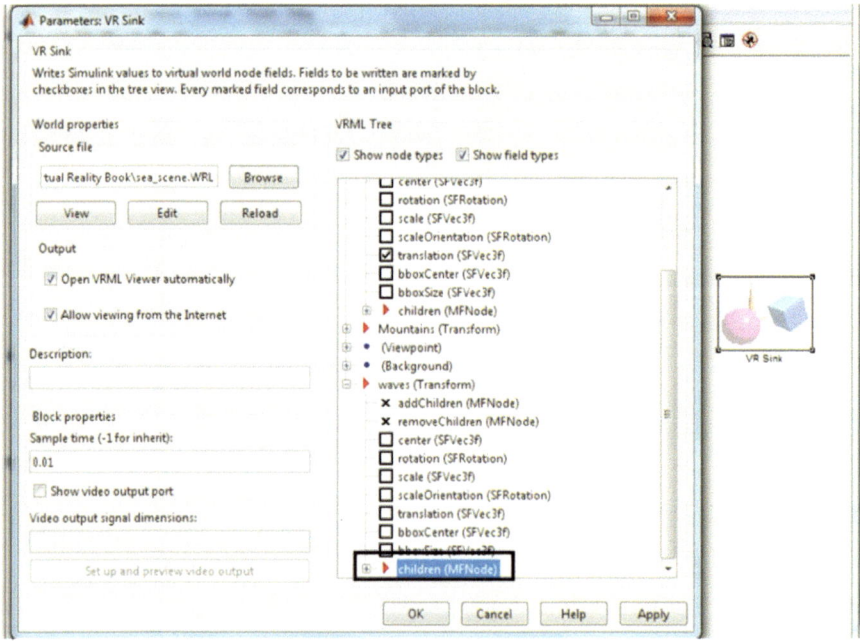

Fig. 12.8   Children of waves is shown in the rectangle

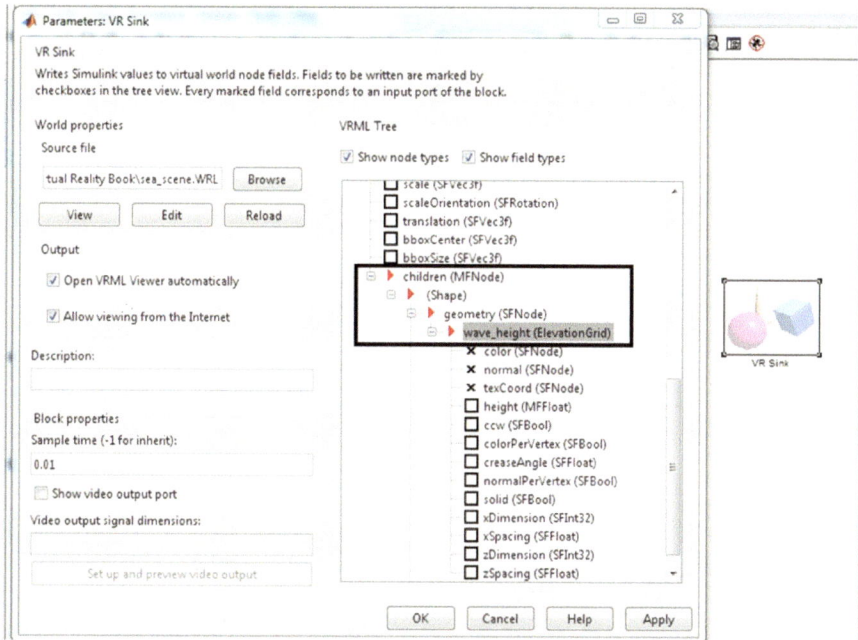

Fig. 12.9 (Shape), geometry, and wave_height are shown in the rectangle

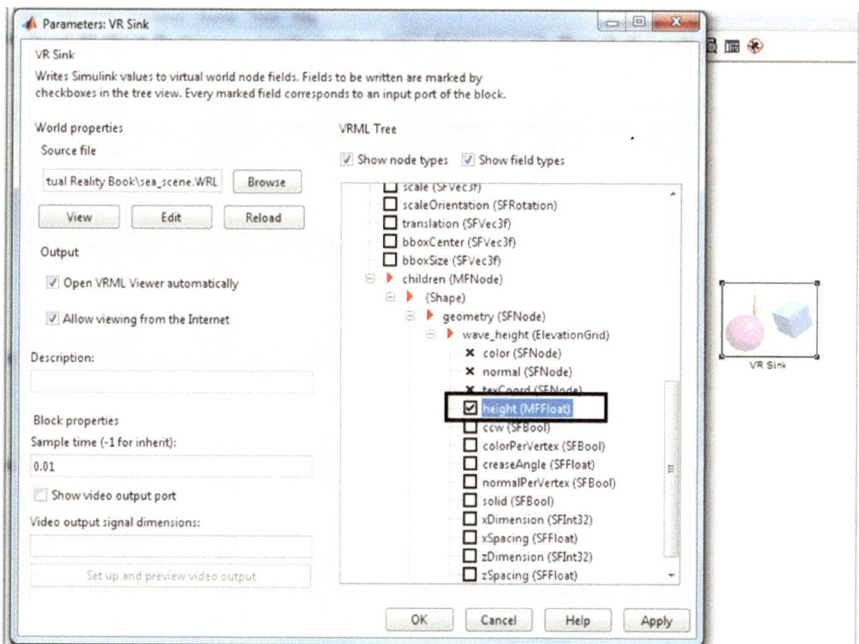

Fig. 12.10 The square to the left of height property is shown in the rectangle

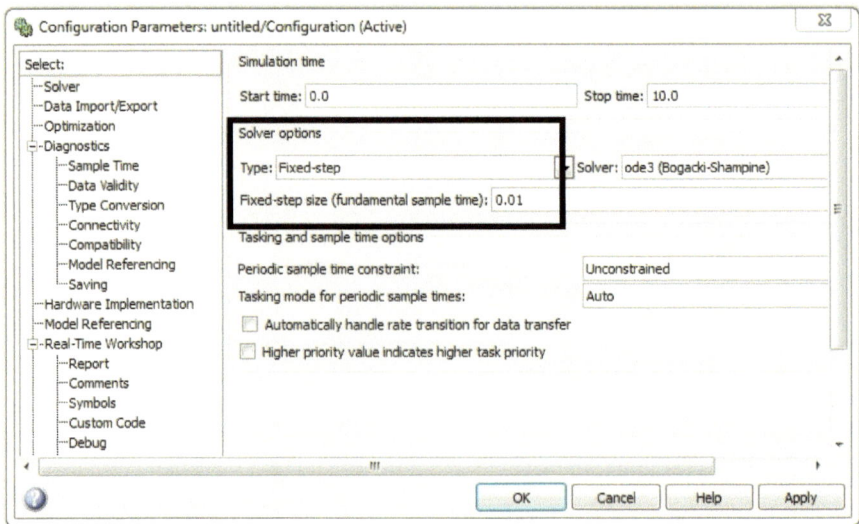

Fig. 12.11 Solver options

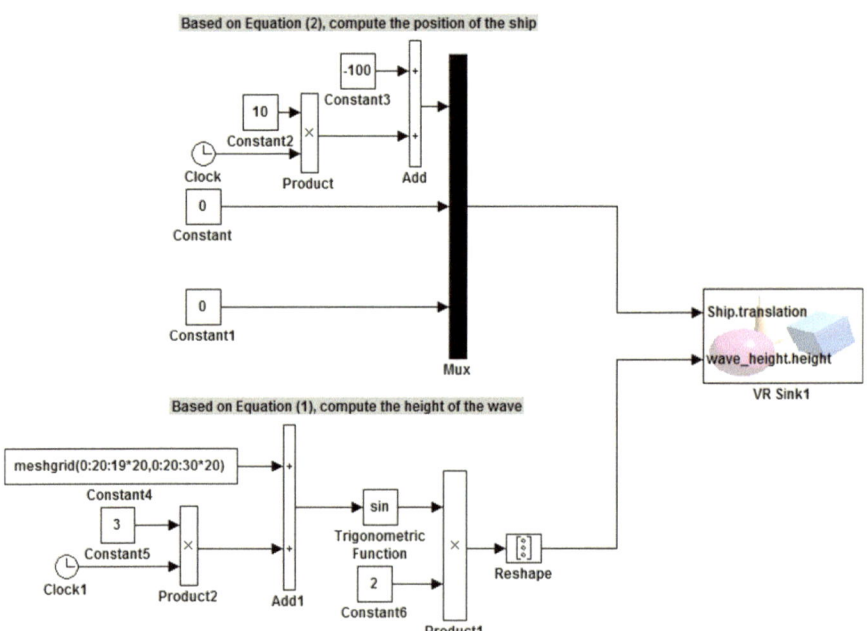

Fig. 12.12 Final model

To run the Simulink® model, double-click the **VR Sink** to open the **VRML** viewer. Start the simulation by clicking on the **Start simulation** button either from the **VRML** viewer or from the Simulink® model (Fig. 12.13). A better practice would be to start the simulation from the **VRML** viewer as the reader might want to

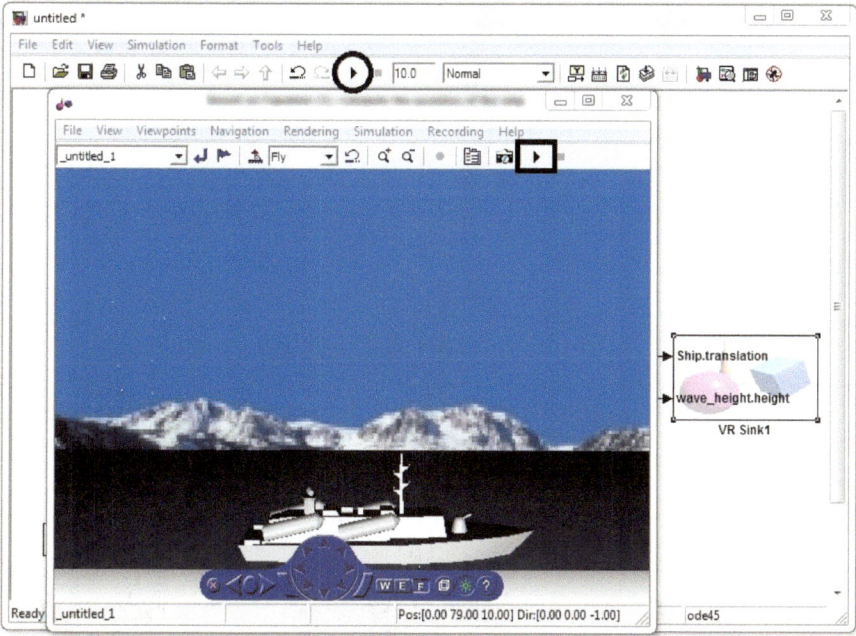

Fig. 12.13 Start Simulation buttons inside the square and circle for VRML viewer and the Simulink® model, respectively

keep the window of the viewer on the top of the model to be able to see the animation. Clicking on the Simulink® model would place the model in front of the **VRML** viewer.

12.4   Application Problem: Ship Heading Controller

The main objective of this section is to design a controller for the ship heading ψ (the angle of the ship with the x-axis in Fig. 12.14) by using the rudder angle α and animate the controlled motion of the ship in the virtual scene by using Simulink® model. The virtual scene that will be used for this problem is *sea_scene2.wrl* that was constructed in the application problem of Chap. 7. The proposed controller for this problem is the fuzzy logic controller. The fuzzy controller has two inputs and one output. The inputs are the normalized error and its time derivative, whereas the output is the normalized control action which yields the angle of the rudder, α, when multiplied by the gain K_{α} (Fig. 12.15). The relation between ship heading ψ and the rudder angle α is given by (Khaled 2010):

$$\ddot{\psi} = \frac{M + \sin(\alpha)}{10} \tag{12.3}$$

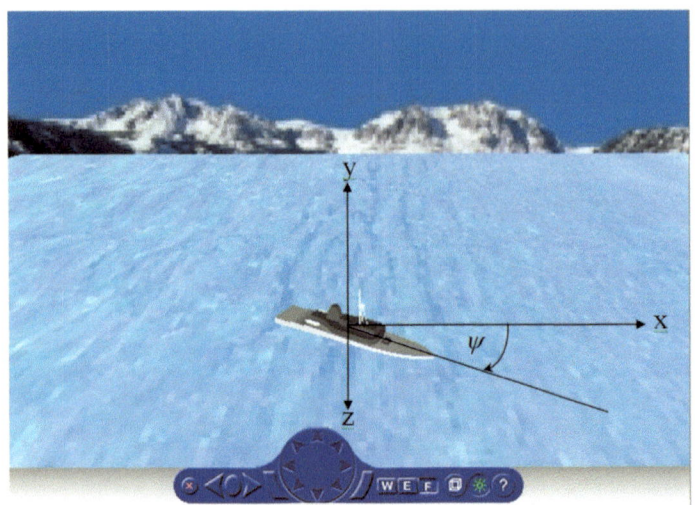

Fig. 12.14 Heading of the ship

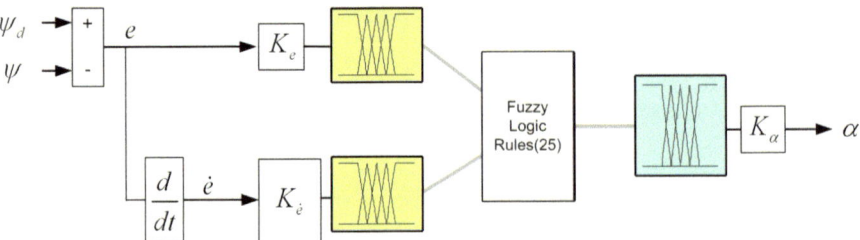

Fig. 12.15 Fuzzy logic heading controller

where M is the sum of the external moments around the y-axis of the ship. For this example, $M = 0.5\sin(3t)$ and $\alpha \in [\frac{-\pi}{5}, \frac{\pi}{5}]$. Note that rudder machines are practically ineffective to control the heading of the ship at low velocities of ships, but to simplify the problem, it was assumed otherwise.

Solve for the following:

1. Using the MATLAB® Fuzzy Logic Toolbox (type in **fuzzy** in the MATLAB® workspace), construct the fuzzy controller based on the rules given in Table 12.1 and the membership functions in Figs. 12.16, 12.17, and 12.18.
2. Using the equation of motion of the heading given in Eq. 12.3, tune the gains for the fuzzy controller (K_e, $K_{\dot{e}}$, and K_α) that will control the ship's heading to a desired heading $\psi_d = \frac{\pi}{3}$ with a steady state error of 0.01 rad starting from zero initial conditions ($\psi = 0$ and $\dot{\psi} = 0$) in 25 s.

Table 12.1 Rules for the heading fuzzy logic controller

$\dot{e}$ \ e	NL	NS	Z	PS	PL
NL	NL	NL	NL	NL	PL
NS	NL	NS	NS	Z	PL
Z	NL	NS	Z	PS	PL
PS	NL	Z	PS	PS	PL
PL	NL	PL	PL	PL	PL

NL stands for negative large, *NS* stands for negative small, *Z* stands for zero, *PS* stands for positive small and *PL* stands for positive large

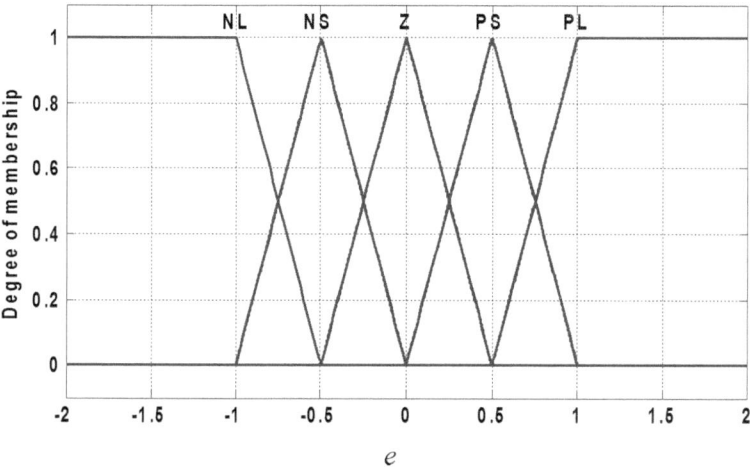

Fig. 12.16 Membership functions for the error e

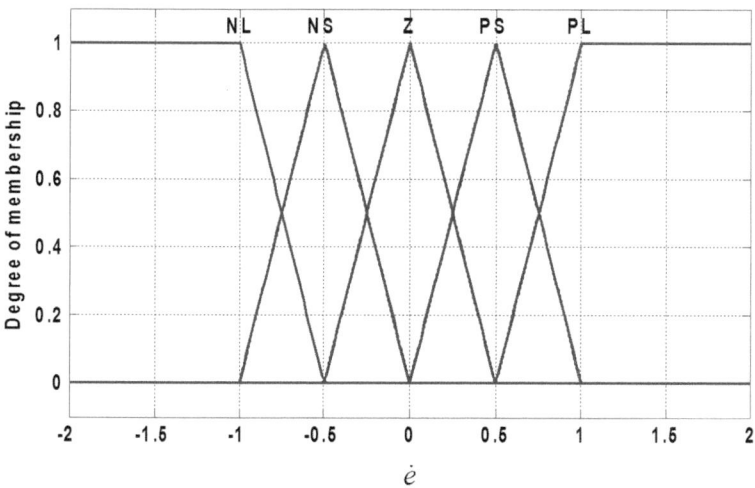

Fig. 12.17 Membership functions for the time derivative of the error $\dot{e}$

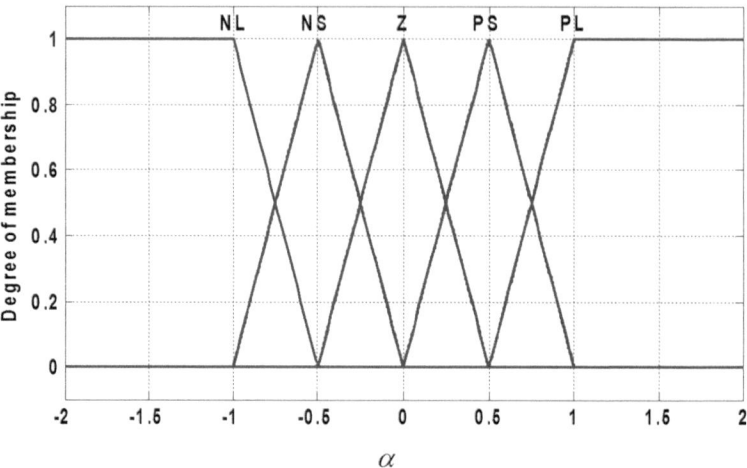

Fig. 12.18 Membership functions for the normalized controller output

3. Develop the Simulink® model that will:

- Animate the sea waves of *sea_scene2.wrl* based on Eq. 12.1.
- Implement the developed fuzzy logic controller for the heading to compute the rudder angle α.
- Compute the heading of the ship based on the rudder angle provided by the controller and using the equation of motion of the heading given by Eq. 12.3.
- Change **rotation** property of **Patrol Boat** based on the angle ψ. (Hint: Rotate the **Patrol Boat** around the y-axis by an angle ψ.)

The solution of the problem can be downloaded from Springer's web site http:// extras.springer.com/. The files can be found in folder /Chapter 12/Application Problem.

References

Books

Khaled N (2010) Guidance and control of ships: modeling, estimation, guidance and control. VDM Verlag Dr. Müller, Saarbrücken

Perez T (2005) Ship motion control: course keeping and roll stabilisation using rudder and fins. Springer, New York

Index

N. Khaled, *Virtual Reality and Animation for MATLAB® and Simulink® Users:*
Visualization of Dynamic Models and Control Simulations,
DOI 10.1007/978-1-4471-2330-9_9, © Springer-Verlag London Limited 2012